D0108621

National Academy Press

The National Academy Press was created by the National Academy of Sciences to publish the reports issued by the Academy and by the National Academy of Engineering, the Institute of Medicine, and the National Research Council, all operating under the charter granted to the National Academy of Sciences by the Congress of the United States.

Prudent Practices for Handling Hazardous Chemicals in Laboratories

Committee on Hazardous Substances in the Laboratory

Assembly of Mathematical and Physical Sciences

National Research Council (U.S.)

National Academy Press
Washington, D.C. 1981

NOTICE: The project that is the subject of this report was approved by the Governing Board of the National Research Council, whose members are drawn from the councils of the National Academy of Sciences, the National Academy of Engineering, and the Institute of Medicine. The members of the committee responsible for the report were chosen for their special competences and with regard for appropriate balance.

This report has been reviewed by a group other than the authors according to procedures approved by a Report Review Committee consisting of members of the National Academy of Sciences, the National Academy of Engineering, and the Institute of Medicine.

The National Research Council was established by the National Academy of Sciences in 1916 to associate the broad community of science and technology with the Academy's purposes of furthering knowledge and of advising the federal government. The Council operates in accordance with general policies determined by the Academy under the authority of its congressional charter of 1863, which establishes the Academy as a private, non-profit, self-governing membership corporation. The Council has become the principal operating agency of both the National Academy of Sciences and the National Academy of Engineering in the conduct of their services to the government, the public, and the scientific and engineering communities. It is administered jointly by both Academies and the Institute of Medicine. The Academy of Engineering and the Institute of Medicine were established in 1964 and 1970, respectively, under the charter of the Academy of Sciences.

This material is based on work supported by the National Science Foundation under Grant No. PRM-7916766, the National Institutes of Health under Agreement No. 1Y-1DS00001-00, the Environmental Protection Agency under Agreement No. EPA-80-D-X0979, the Alfred P. Sloan Foundation under Grant-in-Aid No. B1979-1, the American Chemical Society, and the Chemical Manufacturers Association.

Library of Congress Cataloging in Publication Data

National Research Council. Committee on Hazardous Substances in the Laboratory.
Prudent practices for handling hazardous chemicals in laboratories.

Bibliography: p.
Includes index.
1. Chemical laboratories—Safety measures.
I. Title.
QD51.N32 1980 542'.028'9 80-26877

Available from
NATIONAL ACADEMY PRESS
2101 Constitution Avenue, N.W.
Washington, D.C. 20418

Printed in the United States of America

COMMITTEE ON HAZARDOUS SUBSTANCES IN THE LABORATORY

HERBERT O. HOUSE, Georgia Institute of Technology, *Chairman*
ROBERT A. ALBERTY, Massachusetts Institute of Technology
JEROME A. BERSON, Yale University
ROBERT W. DAY, University of Washington
THOMAS S. ELY, Eastman Kodak Company
RONALD W. ESTABROOK, University of Texas Health Science Center
ANNA J. HARRISON, Mount Holyoke College
DONALD M. JERINA, National Institute of Arthritis, Metabolism & Digestive Diseases
MARVIN KUSCHNER, State University of New York Health Science Center
ELIZABETH C. MILLER, University of Wisconsin
ROBERT A. NEAL, Vanderbilt University
J. E. RALL, National Institute of Arthritis, Metabolism & Digestive Diseases
GEORGE ROUSH, JR., Monsanto Company
ALFRED W. SHAW, Shell Development Company
HOWARD E. SIMMONS, E.I. duPont de Nemours & Company, Inc.

Subcommittee to Define Hazards of Known Laboratory Chemicals and to Critically Examine Existing Epidemiological Studies

JEROME A. BERSON, Yale University, *Chairman*
ROBERT W. DAY, University of Washington
LEON GORDIS, Johns Hopkins School of Public Health
DONALD M. JERINA, National Institute of Arthritis, Metabolism & Digestive Diseases
BLAINE C. MCKUSICK, E.I. du Pont de Nemours & Company, Inc.
ELIZABETH C. MILLER, University of Wisconsin
GEORGE ROUSH, JR., Monsanto Company

Subcommittee to Recomment Procedures for Procurement, Storage, Distribution, and Disposal of Chemicals

ROBERT A. ALBERTY, Massachusetts Institute of Technology, *Chairman*
RONALD W. ESTABROOK, University of Texas Health Service Center
JOHN M. FRESINA, Massachusetts Institute of Technology
ANNA J. HARRISON, Mount Holyoke College
WILLIAM G. MIKELL, E.I. du Pont de Nemours & Co., Inc.

ROBERT A. NEAL, Vanderbilt University
MAX TISHLER, Wesleyan University

Subcommittee to Develop a Guide for Work Practices Related to the Use of Chemicals in Laboratories

HERBERT O. HOUSE, Georgia Institute of Technology, *Chairman*
THOMAS S. ELY, Eastman Kodak Company
CLAYTON H. HEATHCOCK, University of California, Berkeley
RALPH R. LANGNER, Dow Chemical Company
MARVIN KUSCHNER, State University of New York Health Sciences Center
ALFRED W. SHAW, Shell Development Company
HOWARD E. SIMMONS, E.I. du Pont de Nemours & Company, Inc.
RALPH G. SMITH, University of Michigan

Assembly of Mathematical and Physical Sciences Liaison Representatives

ELKAN BLOUT, Harvard University Medical School
THEODORE L. CAIRNS, Greenville, Delaware
WILLIAM G. DAUBEN, University of California, Berkeley

NRC Staff

WILLIAM SPINDEL, *Study Director*
FRANCES R. ZWANZIG, *Staff Officer*
MICHAEL M. KING, *Consultant*
MARTIN PAUL, *Consultant*
PEGGY J. POSEY, *Administrative Assistant*
JEAN E. YATES, *Secretary*

Contents

List of Tables

List of Figures

Figure

Preface

In December 1975, an *ad hoc* planning group from the Assembly of Mathematical and Physical Sciences of the National Academy of Sciences concluded that there was need for a National Research Council study to recommend procedures for the safe handling and disposal of toxic substances in laboratories. After further discussion, the objective was defined as the development of

1. a definition of the problem, including an evaluation of the available epidemiological evidence, and
2. a set of recommended procedures for handling and disposing of hazardous substances in laboratories.

The study was to encompass all types of laboratories where chemicals are used and to include not only hazards arising from the acute or chronic toxicity (including carcinogenicity) of chemicals but also such physical hazards as fires or explosions that can arise from handling chemicals. However, the study was not intended to include hazards that might exist in pilot plant or manufacturing facilities nor be concerned with physical hazards (e.g., radioactivity, lasers) or biological hazards (e.g., pathogenic bacteria, virus infections, recombinant DNA) that may exist in a laboratory. Therefore, the recommendations and procedures outlined in this report deal specifically with chemistry and life sciences laboratories where relatively small quantities of chemicals are used and may differ from those prescribed for dealing with large-scale continuing chemical processes such as those found in an industrial setting.

With this objective in mind, funding for the study was obtained and a parent committee and three subcommittees were assembled. The committee and its subcommittees were intended to represent the concerns of both chemical and life science laboratories and the various types of workers in academic, governmental, and industrial laboratories. The first phase of the study was a survey of the safety practices currently being followed by representative groups of academic and industrial laboratories. The combined experience and knowledge of individuals in these various types of laboratories have served as the basis for the specific procedures recommended in later sections of this report. In a further effort to obtain the viewpoints of the variety of laboratory workers for whom this report was written, draft copies have been distributed to a number of laboratory scientists and others concerned with the health and safety of laboratory workers. The suggestions and comments obtained from this group of individuals were considered in preparing the final report.

Prudent Practices for Handling Hazardous Chemicals in Laboratories

Overview and General Recommendations

1. NATURE AND SCOPE OF LABORATORY HAZARDS

People who work in scientific laboratories are exposed to many kinds of hazards. This can be said of most workplaces; in some, the hazards are well recognized—those of ordinary fire, for example—and the precautions to be taken are obvious. Laboratories, however, involve a greater variety of possible hazards than do most workplaces, and some of those hazards call for precautions not ordinarily encountered elsewhere. In particular, laboratories in which chemicals are used must be prepared to deal with substances known to be hazardous, with the possible hazards of new substances, and with new types of experiments. However, in contrast to manufacturing plants, laboratories usually handle only small amounts of materials and exposure to a particular material seldom extends over a protracted time. In this respect, industrial laboratories differ little from university, governmental, or other research laboratories. Colleges and universities have the special responsibility of administering instructional laboratories, where relatively inexperienced students must be introduced to the safety precautions necessary to the conduct of various laboratory operations.

Safety has always been a concern of those working with hazardous materials in the laboratory. (See, for example, *Laboratory Planning for Chemistry and Chemical Engineering*, Lewis, H. F., Ed.; Reinhold: New York, 1962, a publication sponsored by the National Research Council, which includes an entire section on health and safety factors as well as specific design applications to such needs as those of radioactivity laboratories, hospital research laboratories, and high-pressure laboratories.)

Few laboratory chemicals are without hazards of various kinds and degrees. A substantial part of the time spent by students in instructional laboratories is used in learning how to handle the materials and conduct routine operations with them safely. In research, by its very nature, one may be working with new chemicals or with biologically active materials presenting unknown hazards. However, the past 30 years have witnessed extraordinary changes in the conduct of chemical observations, particularly in the ability to work with ever more minute samples and the use of automated instrumentation. Furthermore, our understanding of environmental and health problems has increased steadily, and this is being reflected in improvements in the design of and working conditions in laboratories.

This report is the product of a study initiated by research investigators active in the fields of chemistry and the life sciences. It deals with the hazards of using chemicals in the laboratory and with certain related hazards. Its purpose is to assemble information, derived mainly from critical assessment of the best current knowledge and practice, that will be useful to laboratory workers and administrators in designing and improving safety plans appropriate to their needs.

The hazards of handling chemicals in the laboratory may be classified broadly as physical or chemical. Physical hazards include those of fire, explosion, and electric shock, which are extremely serious and not unfamiliar in most laboratories. Other physical hazards arise from means of containment, such as cylinders of compressed gases, cryogenic equipment, furnaces, refrigerators, and glass apparatus.

Chemical hazards are associated with their toxic effects and may be subclassified as acute or chronic. Acute hazards are those capable of producing prompt or only slightly delayed effects (such as serious burns, inflammation, allergic responses, or damage to the eyes, lungs, or nervous system). Some chemicals are extremely dangerous in this respect, and small amounts can cause death or severe injury very quickly. Some, such as chlorine or ammonia, give considerable warning, but others, such as carbon monoxide, are not readily detected and consequently are more insidious.

Other toxicological effects of chemicals may be delayed or develop only after exposure over long periods of time and are referred to as chronic. These effects may involve cumulative damage to many different organs or parts of the body. Some are reversed on elimination of exposure to the chemical, but some are nearly irreversible, especially after much damage has occurred. Carcinogenic effects are usually chronic effects. In recent years, we have become increasingly conscious of the variety and seriousness of all these chronic effects, which are insidious because of the long delay in their appearance. More research is needed to develop a better understanding of how these effects are produced and what can be done to prevent them.

The identification and regulation of carcinogens in particular is currently receiving much attention. Although fewer than two dozen chemicals are definitely known to cause cancer in humans, several hundred can do so in laboratory animals. In addition, substances known to be mutagenic or DNA-damaging in *in vitro* systems are suspected to be carcinogenic. The number of substances suspected to be carcinogenic is 1000-2000; the large range of uncertainty is due to the limited number of validated experimental studies providing data for many of these substances.

Although there is currently a justifiably strong interest in chronic effects, particularly carcinogenesis, we believe that acutely toxic, explosive, and flammable substances constitute at least equally important and immediate risks to laboratory workers. In addition, precautions that reduce exposure to acutely hazardous substances also reduce the probability that chronic effects will be incurred. The procedures recommended in this report reflect this view of laboratory safety.

Much chemical research is concerned with new molecular structures. It is common to synthesize substances that have never existed. Such research is carried out in both universities and industry and is a major source of innovation in both fundamental and applied chemistry. Although the biological activity of a substance can sometimes be inferred from knowledge of its structure, most often it cannot. Nevertheless, when dealing with new structures, the laboratory worker must attempt to anticipate, by analogy to related structures of known toxicity, when a new substance may have exceptionally high acute or chronic toxicity. This forethought should be an important part of planning all research involving new chemicals. However, until actual toxicological data are available, all new substances should be handled as though they were toxic.

Finally, this report is concerned not only with the hazards that are encountered in the laboratory itself but also with the related hazards involved in handling laboratory chemicals on the loading dock and in storerooms and stockrooms, in transporting them, and in disposing of them. Because many other persons may be exposed to chemical hazards arising from laboratory operations, it is necessary to take actions to warn them of and protect them from such hazards. These persons include maintenance personnel and others who may be in the laboratory infrequently or may be exposed to hazards in case of an accident in a laboratory or during transportation.

2. REGULATION OF USE OF CHEMICALS

Several types of regulations affect the use of chemicals in the laboratory and the related operations of transportation and disposal of chemicals. (For example, the handling of ethanol and certain pharmacological agents has been controlled for some time. These regulations are not discussed here.)

The Occupational Safety and Health Administration (OSHA), U.S. Department of Labor (DOL), has promulgated a number of regulations that limit the exposure of workers to chemicals. The OSHA regulations are permissible exposure limits (PELs) based on the Threshold Limit Values (TLVs®) that have been developed by the American Conference of Governmental Industrial Hygienists for about 500 substances (The 1980 TLVs are appended to this report as Appendix B). In general, these are levels that are not to be exceeded by the average exposure over an 8-hour working day. In addition, OSHA has classified (29 CFR 1910; compare also 45 Fed. Reg. 5002-5296) certain substances as carcinogens and requires that exposure to these substances be controlled, certain records be kept, certain warnings be given, and medical examinations be made available to persons who may be exposed. OSHA regulations apply to almost all workplaces (the exceptions include state and local agencies, workplaces in states in which a state organization has the responsibility for regulating working conditions, and federal agencies other than the DOL). OSHA is continuing the process of developing additional regulations for carcinogens, and the regulatory situation in this area is expected to change frequently.

The U.S. Environmental Protection Agency (EPA) has the responsibility for regulation of chemicals in the air, in water, and on land. These regulations have an important effect on the disposal of chemicals. In addition, there are state and local regulations that place limits on what can be discharged to the sewer system or the atmosphere or put in waste disposal areas. New regulations to control chemical wastes are being developed under the *Resource Conservation and Recovery Act of 1976* (RCRA); this is an area where increased regulation can be expected (see 45 Fed. Reg. 12,722-12,754).

The Laboratory Chemical Carcinogen Safety Standards Subcommittee of the Committee to Coordinate Environmental and Related Problems, U.S. Department of Health and Human Services, is currently developing a draft of guidelines for the laboratory use of chemical substances posing a potential occupational carcinogenic risk.

Finally, the U.S. Department of Transportation (DOT) controls the shipping of chemicals and their transport on any type of public facility, including a private vehicle using public roads. These regulations also affect the transport of chemicals as part of a disposal process.

3. GOALS OF A LABORATORY SAFETY PROGRAM

An important objective of this report is to provide guidance to all laboratory workers who use chemicals, so that they can perform their work safely. Experience has shown, especially in industry, that the laboratory can be a safe workplace. This record, however, has been achieved only through vigorous safety programs. The goals of a laboratory safety plan should be to protect from injury those working in the laboratory, others who may be exposed to hazards from the laboratory, and the environment.

Laboratory Workers and Students in Instructional Laboratories

Each individual working in a laboratory should be informed about safety in connection with that particular laboratory and the work going on there. This includes all research staff, faculty, postdoctoral fellows, technicians, teaching assistants, and students. No responsible investigator would knowingly conduct research in a manner that jeopardized anyone's health or safety.

Support Personnel

The stockroom personnel, maintenance personnel, technical assistants, animal care personnel, persons transporting chemicals, and others in the vicinity of the laboratory may also be exposed to potential physical and chemical hazards in connection with work going on in the laboratory. They should be informed about the risks involved and educated about how to avoid potential hazards and what to do in the event of an accident.

The Environment

Chemicals must be disposed of in such a way that people, other living organisms, and the environment generally are subjected to minimal harm by the substances used or produced in the laboratory. Both the laboratory workers and the supporting personnel should know and use acceptable disposal methods for various chemicals.

4. RESPONSIBILITY FOR LABORATORY SAFETY

First and foremost, the protection of health and safety is a moral obligation. An expanding array of federal, state, and local laws and regulations makes it a legal requirement and an economic necessity as well. In the final analysis, laboratory safety can be achieved only by the exercise of judgment by informed, responsible individuals. It is an essential part of the development of scientists that they learn to work with and to accept the responsibility for the appropriate use of hazardous substances.

Liability for a laboratory misadventure (accident, illness, environmental damage) may lie with the individual experimenters, their immediate supervisors, other officers of the institution, or the institution itself, depending on the circumstances and applicable laws—federal, state, and local. In view of the small number of cases decided, it is impossible to predict the outcome of a law suit in this area. Each institution, therefore, should seek expert legal advice pertinent to its particular situation, so that potential liability can be estimated ahead of time if possible.

Past experience has shown that voluntary safety programs are often inadequate. Good laboratory practice requires mandatory safety rules and programs. To achieve safe conditions for the laboratory worker, a program must include (a) regular safety inspections at intervals of no more than 3 months (and at shorter intervals for certain types of equipment, such as eyewash fountains), (b) disposal procedures that ensure disposal of waste chemicals at regular intervals, (c) formal and regular safety programs that ensure that at least some of the full-time personnel are trained in the proper use of emergency equipment and procedures, and (d) regular monitoring of the performance of ventilation systems.

A sound safety organization that is respected by all is essential; a good laboratory safety program must always be based on participation of both laboratory administration and students and/or employees. Laboratory workers and their institutions or companies are strongly encouraged to follow the safety practices recommended in this report. Large industrial organizations often have safety programs costing millions of dollars annually. Academic institutions have been rather casual about safety programs; the time and money devoted to safety programs by academic institutions will have to increase. University faculties and administrations need to develop, support, and enforce well-defined safety policies. The importance of a safety-minded point of view among all employees or students must be inculcated by the institution. In the end, the individual worker must learn to think about possible hazards and seek information and advice before beginning any experiment.

The ultimate responsibility for safety within an institution lies with its chief executive officer. This individual must ensure that an effective institutional safety program is in place. The chief executive officer and all immediate associates (vice presidents, deans, department heads, and such) must exhibit a sincere and continuing interest in the safety program, and this interest must be obvious to all. An excellent safety program that is ignored by top management (until after an accident) will certainly be ignored by everyone else. Essential to an effective institutional safety program is an institutional safety coordinator (or institutional safety officer). This individual should have appropriate training and be qualified in those areas of safety that are relevant to the activities of the institution. Records should document that the facilities available and the precautions taken in carrying out activities of the institutions are compatible with the current state of knowledge of the potential risks and the law. Experimental work involving chemicals is a subset of those activities.

The responsibility for safety in a department (or other administrative unit) lies with its chairperson or supervisor. Essential to an effective departmental safety program is a departmental safety coordinator (or departmental safety officer). In smaller institutions, it may be possible for one person to perform more than one set of duties. For example, a significant fraction of the time of a departmental faculty member might be allotted to the duties of departmental safety coordinator. However, it must be recognized that such duties are time-consuming and will require regular attention.

The responsibility for the authorization of a specific operation, the delineation of the appropriate safety procedures, and the instruction of those who will carry out the operation lies with the project director of that undertaking.

The responsibility for safety during the execution of an operation lies with the operator(s) executing that operation. Operators frequently include workers, technicians, and students. Nevertheless, the primary responsibility remains with the project director.

A typical pattern of interactions among these various individuals is shown in Figure 1.

An effective departmental safety coordinator must be committed to the attainment of a high level of safety and must work with administrators and investigators to develop and implement policies and practices appropriate for safe laboratory work. In these activities, the safety coordinator requires the cooperation of everyone—workers, technicians, and students, as well as scientists. Collectively, this group must routinely monitor current

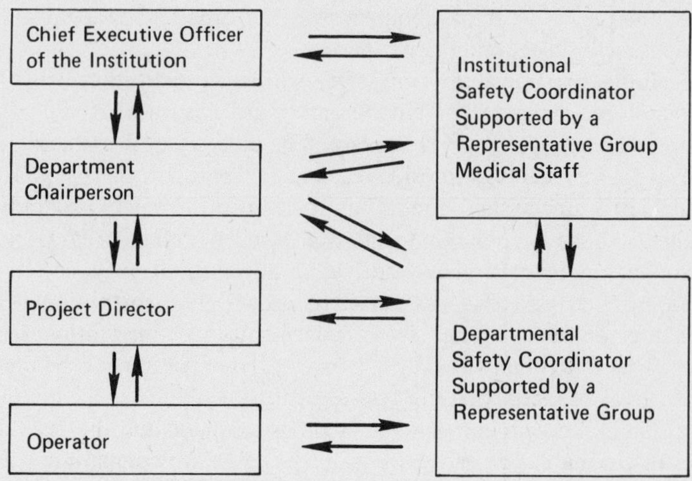

FIGURE 1 Pattern of interactions.

operations and practices, see that appropriate audits are maintained, and seek ways to improve the safety program. If the goals of the laboratory dictate specific operations and the use of specific substances not appropriate to the existing facilities, it is the responsibility of the safety coordinator and the representative group to assist the investigator in acquiring adequate facilities and developing appropriate guidelines.

All accidents and near accidents should be carefully analyzed, and the results of such analyses and the recommendations for the prevention of similar occurrences should be distributed to all who might benefit. This is not to be aimed at fixing blame but at contributing to a safer environment.

The processes involved in the procurement, use, and disposal of chemicals are summarized in Figure 2. The safety coordinator and the representative group must monitor these processes and make provision for orderly disposal of the material should circumstances such as spills or accumulation of unusable or hazardous material dictate such action.

Laboratories that use chemicals have diverse goals: the extension of knowledge about single substances and complex systems (research), the development of goods and services (development), and the extension of intellectual competencies and experimental capabilities of individuals (education). The use and distribution of chemicals must be consistent with the goals of the laboratory. Teaching laboratories and research laboratories have different goals and, therefore, different operating conditions. Although the work done in teaching laboratories can usually be designed

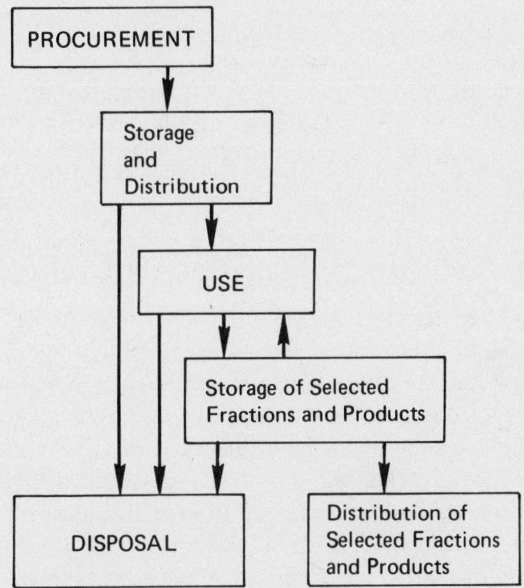

FIGURE 2 Processes involved in the procurement, use, and disposal of chemicals.

to use chemical substances that have well-known properties, work in research laboratories frequently involves chemical substances that have completely unknown properties. Teaching laboratories often involve large numbers of relatively inexperienced students, while research laboratories usually involve a small number of experienced investigators assisted by technicians.

The risk associated with the possession and use of each specific substance is dependent on the following:

1. the knowledge of and commitment to safe laboratory practices of all who handle it;
2. its physical, chemical, and biological properties;
3. the quantity received and the manner in which it is stored and distributed;
4. the manner in which it is used;
5. the manner of disposal of the substance and its derivatives;
6. the length of time it is on the premises; and
7. the number of persons who work in the area and have open access to the substance.

The decision to procure a specific quantity of a specific chemical is a commitment to handle it responsibly from receipt to ultimate disposal. Each operation in which it is handled and each period between operations presents opportunities for misadventure.

5. HANDLING CHEMICALS IN THE LABORATORY

Chemicals occur in almost limitless (and ever-increasing) varieties. For this reason, general precautions for handling almost all chemicals are needed, rather than specific guidelines for each chemical. Otherwise, laboratory work will be needlessly handicapped, practically and economically, by attempts to adhere to a labyrinth of separate guidelines or, more likely, the laboratory worker will simply ignore the entire complex set of guidelines and, consequently, be exposed to excessive hazard.

In subsequent sections of this report, we have taken the viewpoint that, under some circumstances, all chemicals can be hazardous. Accordingly, we have recommended general procedures, applicable to all chemicals, that are designed to minimize the exposure of the laboratory worker to any chemical. In addition, we have noted other precautions that are appropriate in working with substances that are known to be flammable, explosive, or unusually toxic.

It is imperative that the work occurring in teaching and experimental research laboratories be differentiated from that in pilot plants and industrial manufacture. Research in academic and industrial laboratories is carried out on a small scale and, hence, generally involves low levels of exposure of laboratory workers to chemicals. This is particularly true when the laboratory worker makes proper use of the hoods, protective apparel, and other safety devices that should be present in a well-equipped laboratory. Furthermore, in contrast to the typical industrial plant, where workers may be exposed to a limited number of substances over very long periods, the research worker is exposed to a large variety of substances at low levels for brief periods of time. Finally, the professional expertise, common sense, judgment, and safety awareness of the research worker performing chemical operations in the research laboratory most often puts him or her in the best position to judge necessary safety precautions.

Careful attention must be paid to the appropriateness of the experimental work conducted in relation to the adequacy of the physical facilities available and the personnel involved. Once these are established, it is the role of the safety coordinator and the representative group to assist in the development of adequate guidelines for operations. For example, the ventilation facilities available in a given laboratory may preclude certain kinds of work or the use of certain materials.

A continuing appraisal of safety facilities (hoods, incinerators, and such) should be made, and modernization should be instituted whenever they are judged inadequate for the work planned.

6. RECOMMENDATIONS

Safety Plans

Organizations administering laboratories should have safety plans. The administration should actively support safety by ensuring that a safety plan is developed and followed. All persons in the organization must understand their responsibilities and should take appropriate actions to ensure safe operations. In many organizations, safety committees representing the various types of persons exposed to potential hazards are needed to discuss problems, recommend safety measures, and facilitate communication. Most organizations should have safety coordinators to work on problems, to serve as consultants on safety matters, and to support those involved in providing for safe operations. The safety program should be a regular, continuing effort and not merely a stand-by activity that functions for a short time after each laboratory accident.

Medical Program

Any person whose work involves regular and frequent handling of toxicologically significant quantities of material that is acutely or chronically toxic should consult a qualified physician to determine whether it is desirable to establish a regular schedule of medical surveillance. It can be very useful to monitor body concentrations of chronic toxins such as lead or mercury compounds. Also, it is possible that, in the future, biological monitoring (e.g., of certain enzyme levels) may provide an indication of excessive exposure to some toxins and also an indication that some persons are predisposed to above-average sensitivity to certain toxic substances.

It is not clear that a schedule of regular medical surveillance will offer significant benefit to a person working in the laboratory with carcinogens (see *Handling Chemical Carcinogens in the Laboratory: Problems of Safety*, Montesano, R., *et al.*, Eds.; International Agency for Research on Cancer, Lyon, France, 1979, p. 23). Often, the analyses that could provide useful information for medical surveillance have yet to be developed. We therefore recommend that the need for regular medical surveillance be decided on an individual basis by consultation between the laboratory worker and a qualified physician. Medical records, or copies thereof, should be retained by the institution in accord with state and federal regulations. Local, state, or federal regulations sometimes require medical surveillance for specific compounds.

Facilities

The facilities available are an important part of the provision for safe laboratory operations; their capacity should not be exceeded. The facilities should include hoods, an appropriate ventilation system, stockrooms and storerooms, safety equipment, and arrangements for disposal. The performance of the laboratory ventilation system and other safety facilities should be monitored at regular intervals; we recommend at least once every three months. In cases where the facilities are inadequate for the work being done, they should be improved so that they are adequate or the experimental work should be changed so that the safety capacity of the facilities is not exceeded.

Although the energy costs of ventilation, often substantial, are increasing, considerations of economy should never take precedence over ensuring that laboratories have adequate ventilation. Any change in the overall ventilation system to conserve energy should be instituted only after thorough testing of its effects has demonstrated that the laboratory workers will continue to have adequate protection from hazardous concentrations of airborne toxic substances. An inadequate ventilation system can cause increased risk because it can give a false sense of security.

Monitoring of Chemical Substances

For most laboratory environments, we believe that regular monitoring of the airborne concentrations of a variety of different toxic materials is both unjustified and impractical. If care is taken to ensure that (1) the ventilation system (including the hood) is performing and being used properly, (2) the laboratory workers are using proper protective clothing to avoid skin contact, and (3) the laboratory workers are following good hygiene and laboratory safety practices, then even highly toxic materials can be handled without undue hazard.

We believe that there are two circumstances where monitoring of individual compounds is appropriate. In testing or redesigning the hoods and other local ventilation devices in a laboratory, it is often helpful to release a substance (e.g., Freon 11®, sulfur hexafluoride) whose airborne concentration is readily monitored by commercial instruments. Alternatively, laboratory workers can wear personal air-sampling devices to provide a measure of the airborne concentration of some substance in their environment.

If a specific substance that is highly toxic is regularly and continuously used (e.g., three times a week), then instrumental monitoring of that substance may be appropriate. This is especially true if a relatively large amount of the material is being stored or used in the laboratory.

Academic Teaching Laboratories

In general, the students and instructors in academic teaching laboratories should follow the safety procedures recommended for full-time laboratory workers in research and development laboratories. The need for using appropriate protective apparel (safety glasses, gloves, and such), for following general safe laboratory practices, and for providing emergency safety equipment (safety showers, eyewash stations, fire extinguishers, and such) is probably even greater in instructional laboratories where sizable numbers of relatively untrained laboratory workers may be present in relatively close quarters.

The most severe limitation on protective equipment in instructional laboratories is usually the general laboratory ventilation and, especially, auxiliary local exhaust ventilation (hoods or their equivalent). It is unlikely that most academic institutions will be able to provide a laboratory hood for every two students in their instructional laboratories. Therefore, we recommend that the work done and the chemicals used in any instructional laboratory be adjusted according to the quality of ventilation protection that is available in that laboratory. Unless adequate hood space can be provided, it seems prudent to avoid work with substances whose toxicity has not been studied. The selection of the particular substances to use among those whose toxicological properties are known should be based on the quality of ventilation system available.

The PELs of OSHA and the current threshold limit value (TLVs®; see Section I.B.1 and Appendix B) provide useful guides. All work in instructional laboratories should be carried out in such a way that the concentration of each substance being used does not exceed its PEL or recommended TLV. This may be achieved by a combination of experimental design and laboratory ventilation. In general, use of a hood or some equivalent form of local ventilation is desirable when working with any appreciably volatile substance having a TLV of less than 50 ppm. Of course, this generalization is intended to serve as no more than a rough rule of thumb; many substances having higher limit values can pose hazards if they are used without proper planning and precautions. Furthermore, the overall ventilation system in each instructional laboratory should be evaluated at regular intervals, and some monitoring of concentration levels for specific substances may be required in questionable cases.

Disposal of Waste Substances

Some laboratories currently dispose of waste substances by pouring them down the drain or by placing them in drums to be buried in a landfill by an outside contractor. Such indiscriminate disposal is unacceptable and is being curtailed by a combination of local, state, and federal regulations. It is important that an institutional safety plan provide for the regular disposal of waste chemicals. Waste from individual laboratories should be removed at intervals of not more than 1 week to a central waste disposal storage area and then removed from that area at regular intervals. The most practical alternative for removal of combustible material is to construct or contract for access to an incinerator that is capable of incinerating chemical and biological waste materials in an environmentally acceptable manner. The institutional plan for this type of disposal must include consideration of what materials can be incinerated, how they are to be collected and stored, and their mode of transport to the incinerator.

Literature and Consulting Advice

Literature (see Sections I.E.4, I.A.18, and I.H.7) and consulting advice on laboratory safety and on the physical and biological hazards of chemicals should be readily available to those responsible for laboratory operations and those actually involved. Laboratory workers should be encouraged to read about the potential hazards of the work going on in their laboratory and to know about the availability of various resources that describe safe operating conditions. This literature should be available in a form that is readily accessible both to those responsible for laboratory operations and to laboratory workers themselves.

Although a substantial number of people who have expertise in laboratory safety are employed by large chemical companies and by private consulting firms, such persons are not often found in academic institutions. Because modifications of certain safety facilities (e.g., ventilation systems, waste disposal systems) can be very expensive, such modifications should not be undertaken until advice has been sought from persons qualified to make recommendations. The alternatives for an academic institution are either to hire an appropriately qualified person as the institutional safety coordinator or to hire appropriate consultants as needed to obtain advice about specific safety problems. Some chemical companies have discussed the possibility of encouraging contact between members of their staff of safety experts and universities to provide information about safety problems that arise in universities. Such interactions would have immediate benefit for universities and could be expected to increase the safety consciousness of students being trained in universities.

Educational Activities

Educational activities should be provided for all persons who may be exposed to potential hazards in connection with laboratory operations. This group includes faculty, students, laboratory supervisors, laboratory workers, maintenance and storeroom personnel, and others who are close to laboratories. New persons coming into the laboratory or related jobs should be educated about safety procedures and the procedures to use in the event of accidents.

These institutional education programs should be regular, continuing activities and not simply once-a-year presentations provided for groups of new students or employees.

There is need for the publication of a regular series of articles dealing with laboratory safety and with new potential laboratory hazards that have been found by laboratory workers. These articles should be written by persons who have expertise and experience in the area of laboratory safety and should be reviewed with care to be certain that the information published (e.g., data concerning acute or chronic toxicity) is based on reliable experimental evidence. We recommend that scientific societies publish a regular series of articles concerning laboratory safety in periodicals that are widely read by laboratory workers. We further recommend that the content of the present report be reproduced in such a way that the recommended safety procedures contained herein can be made available to a large number of laboratory workers at a reasonable price. Because the content of this report will require revision as new laboratory hazards are identified and as more toxicological data become available, we also recommend that new editions of *Prudent Practices for Handling Hazardous Chemicals in Laboratories* be published at regular intervals by an agency that is representative of and responsible to the scientific community. Such new editions should continue, as does the present report, to be applicable to both academic and industrial research laboratories investigating chemistry or the life sciences.

Effects on Health of Chemicals; Biological Hazards

Epidemiological evidence supporting or refuting the finding of higher risk of death from cancer among chemists, in comparison with members of reference or other groups, is equivocal. Until quite recently, the preponderance of studies supported the view that those in the occupational category of chemist, as defined in the separate studies, did have a higher risk of death from cancer. Recent evidence is at variance with the previously reported findings. There is, thus, inconsistency in the results of the various epidemiological studies.

All of the epidemiological studies reported to date can be criticized on many grounds; thus, interpretation of the findings is complicated. For example, there was a lack of consistency among the sites of cancers reported as in excess of expectation for several of the groups studied. Furthermore, methodological difficulties in the design of the several studies compromise ready acceptance of the findings. These studies do not distinguish members of the population groups according to the degree of exposure and, because of the presumed long latency period of cancer, any increased incidence probably represents exposure that occurred some time in the past. These issues and other related reservations are described in greater detail in Appendix A.

However, although the results of the epidemiological investigations are equivocal on the point of whether or not there exists a special risk of death from cancer for those variously classified as chemists, the undeniable hazard of handling a variety of chemicals provides good and sufficient reason for laboratories to bring their operations into compliance with the practices recommended here.

More research is needed on the effects of chemicals on health, including epidemiological studies. We recommend that the available toxicological information on specific substances be evaluated by qualified persons and then rewritten and published at regular intervals in a form that will be useful to laboratory workers handling chemicals. In this report, we illustrate the type of information we have in mind in the form of sample data sheets for 33 compounds. A compilation of such data sheets, revised and augmented at regular intervals under the aegis of an organization responsible to the scientific community, would be a valuable resource to which laboratory workers could turn to learn of any special hazards involved in handling or disposing of a specific substance in the laboratory. Neither the currently available company product data sheets nor the various compilations of toxicological data accessible provide information in a form that is really useful to a person who wishes to work with a given substance in the laboratory.

This report does not, and was not intended to, address various biological hazards that arise in laboratories; thus, procedures for safe handling and disposal of bacterial or viral cultures are not considered. We believe that biological laboratory hazards are clearly worthy of a study of the type represented by this report.

I

PROCEDURES FOR WORKING WITH CHEMICALS IN LABORATORIES

I.A

General Recommendations for Safe Practices in Laboratories

It is impossible to design a set of rules that will cover all possible hazards and occurrences. Some general guidelines are given below that experience has shown to be useful for avoiding accidents or reducing injuries in the laboratory.

The most important rule is that everyone involved in laboratory operations—from the highest administrative level to the individual worker—must be safety minded. Safety awareness can become part of everyone's habits only if the issue of safety is discussed repeatedly and only if senior and responsible staff evince a sincere and continuing interest and are perceived by all their associates as doing so. The individual, however, must accept responsibility for carrying out his or her own work in accordance with good safety practices and should be prepared in advance for possible accidents by knowing what emergency aids are available and how they are to be used.

The supervisor of the laboratory has overall safety responsibility and should provide for regular formal safety and housekeeping inspections (at least quarterly for universities and other organizations that have frequent personnel changes and semiannually for other laboratories) in addition to continual informal inspections. Laboratory supervisors have the responsibility of ensuring that (a) workers know safety rules and follow them, (b) adequate emergency equipment in proper working order is available, (c) training in the use of emergency equipment has been provided, (d) information on special or unusual hazards in nonroutine work has been distributed to the laboratory workers, and (e) an appropriate safety orientation has been given to individuals when they are first assigned to a laboratory space.

The laboratory worker should develop good personal safety habits: (a) eye protection should be worn at all times, (b) exposure to chemicals should be kept to a minimum, and (c) smoking and eating should be avoided in areas where chemicals are present.

Advance planning is one of the best ways to avoid serious incidents. Before performing any chemical operation, the laboratory worker should consider "What would happen if . . . ?" and be prepared to take proper emergency actions.

Overfamiliarity with a particular laboratory operation may result in overlooking or underrating its hazards. This attitude can lead to a false sense of security, which frequently results in carelessness. Every laboratory worker has a basic responsibility to himself or herself and colleagues to plan and execute laboratory operations in a safe manner.

I.A.1 GENERAL PRINCIPLES

Every laboratory worker should observe the following rules:

1. Know the safety rules and procedures that apply to the work that is being done. Determine the potential hazards (e.g., physical, chemical, biological) (see Chapters I.B-D) and appropriate safety precautions before beginning any new operation.
2. Know the location of and how to use the emergency equipment in your area, as well as how to obtain additional help in an emergency, and be familiar with emergency procedures (see Sections I.F.1-3).
3. Know the types of protective equipment available and use the proper type for each job (see Sections I.F.4-7).
4. Be alert to unsafe conditions and actions and call attention to them so that corrections can be made as soon as possible. Someone else's accident can be as dangerous to you as any you might have.
5. Avoid consuming food or beverages or smoking in areas where chemicals are being used or stored (see Section I.A.3).
6. Avoid hazards to the environment by following accepted waste disposal procedures (see Chapters II.E and G). Chemical reactions may require traps or scrubbing devices to prevent the escape of toxic substances.
7. Be certain all chemicals are correctly and clearly labeled. Post warning signs when unusual hazards, such as radiation, laser operations, flammable materials, biological hazards, or other special problems exist.
8. Remain out of the area of a fire or personal injury unless it is your

responsibility to help meet the emergency. Curious bystanders interfere with rescue and emergency personnel and endanger themselves (see Section I.F.5).

9. Avoid distracting or startling any other worker. Practical jokes or horseplay cannot be tolerated at any time.

10. Use equipment only for its designed purpose.

11. Position and clamp reaction apparatus thoughtfully in order to permit manipulation without the need to move the apparatus until the entire reaction is completed. Combine reagents in appropriate order, and avoid adding solids to hot liquids (see Section I.D.3).

12. Think, act, and encourage safety until it becomes a habit.

I.A.2 HEALTH AND HYGIENE

Laboratory workers should observe the following health practices:

1. Wear appropriate eye protection at all times (see Section I.F.1).
2. Use protective apparel, including face shields, gloves, and other special clothing or footwear as needed (see Sections I.F.2 and 3).
3. Confine long hair and loose clothing when in the laboratory (see Section I.F.3).
4. Do not use mouth suction to pipet chemicals or to start a siphon; a pipet bulb or an aspirator should be used to provide vacuum.
5. Avoid exposure to gases, vapors, and aerosols (see Chapter I.H). Use appropriate safety equipment whenever such exposure is likely (see Chapter I.F).
6. Wash well before leaving the laboratory area. However, avoid the use of solvents for washing the skin. They remove the natural protective oils from the skin and can cause irritation and inflammation. In some cases, washing with a solvent might facilitate absorption of a toxic chemical.

I.A.3 FOOD HANDLING

Contamination of food, drink, and smoking materials is a potential route for exposure to toxic substances. Food should be stored, handled, and consumed in an area free of hazardous substances.

1. Well-defined areas should be established for storage and consumption of food and beverages. No food should be stored or consumed outside of this area.
2. Areas where food is permitted should be prominently marked and a

warning sign (e.g., EATING AREA—NO CHEMICALS) posted. No chemicals or chemical equipment should be allowed in such areas.

3. Consumption of food or beverages and smoking should not be permitted in areas where laboratory operations are being carried out.

4. Glassware or utensils that have been used for laboratory operations should never be used to prepare or consume food or beverages. Laboratory refrigerators, ice chests, cold rooms, and such should not be used for food storage; separate equipment should be dedicated to that use and prominently labeled.

I.A.4 HOUSEKEEPING

There is a definite relationship between safety performance and orderliness in the laboratory. When housekeeping standards fall, safety performance inevitably deteriorates. The work area should be kept clean, and chemicals and equipment should be properly labeled and stored.

1. Work areas should be kept clean and free from obstructions. Cleanup should follow the completion of any operation or at the end of each day.
2. Wastes should be deposited in appropriate receptacles.
3. Spilled chemicals should be cleaned up immediately and disposed of properly. Disposal procedures should be established and all laboratory personnel should be informed of them (see Section II.E.6); the effects of other laboratory accidents should also be cleaned up promptly.
4. Unlabeled containers and chemical wastes should be disposed of promptly, by using appropriate procedures (see Chapters II.E and G). Such materials, as well as chemicals that are no longer needed, should not accumulate in the laboratory.
5. Floors should be cleaned regularly; accumulated dust, chromatography adsorbents, and other assorted chemicals pose respiratory hazards.
6. Stairways and hallways should not be used as storage areas.
7. Access to exits, emergency equipment, controls, and such should never be blocked.
8. Equipment and chemicals should be stored properly; clutter should be minimized.

I.A.5 EQUIPMENT MAINTENANCE

Good equipment maintenance is important for safe, efficient operations. Equipment should be inspected and maintained regularly. Servicing schedules will depend on both the possibilities and the consequences of failure. Maintenance plans should include a procedure to ensure that a device that is out of service cannot be restarted.

I.A.6. GUARDING FOR SAFETY

All mechanical equipment should be adequately furnished with guards that prevent access to electrical connections or moving parts (such as the belts and pulleys of a vacuum pump) (see Section I.G.2). Each laboratory worker should inspect equipment before using it to ensure that the guards are in place and functioning.

Careful design of guards is vital. An ineffective guard can be worse than none at all, because it can give a false sense of security. Emergency shutoff devices may be needed, in addition to electrical and mechanical guarding (see Chapter I.G).

I.A.7 SHIELDING FOR SAFETY

Safety shielding should be used for any operation having the potential for explosion such as (a) whenever a reaction is attempted for the first time (small quantities of reactants should be used to minimize hazards), (b) whenever a familiar reaction is carried out on a larger than usual scale (e.g., 5-10 times more material), and (c) whenever operations are carried out under nonambient conditions (see Sections I.D.2 and 3). Shields must be placed so that all personnel in the area are protected from hazard (see Section I.F.1).

I.A.8 GLASSWARE

Accidents involving glassware are a leading cause of laboratory injuries.

1. Careful handling and storage procedures should be used to avoid damaging glassware. Damaged items should be discarded or repaired.
2. Adequate hand protection should be used when inserting glass tubing into rubber stoppers or corks or when placing rubber tubing on glass hose connections. Tubing should be fire polished or rounded and lubricated, and hands should be held close together to limit movement of glass should fracture occur. The use of plastic or metal connectors should be considered.
3. Glass-blowing operations should not be attempted unless proper annealing facilities are available.
4. Vacuum-jacketed glass apparatus should be handled with extreme care to prevent implosions. Equipment such as Dewar flasks should be

taped or shielded (see Section I.D.3). Only glassware designed for vacuum work should be used for that purpose.

5. Hand protection should be used when picking up broken glass. (Small pieces should be swept up with a brush into a dustpan.)

6. Proper instruction should be provided in the use of glass equipment designed for specialized tasks, which can represent unusual risks for the first-time user. (For example, separatory funnels containing volatile solvents can develop considerable pressure during use.)

I.A.9 FLAMMABILITY HAZARDS

Because flammable materials are widely used in laboratory operations (see Chapter I.C), the following rules should be observed:

1. Do not use an open flame to heat a flammable liquid or to carry out a distillation under reduced pressure.
2. Use an open flame only when necessary and extinguish it when it is no longer actually needed.
3. Before lighting a flame, remove all flammable substances from the immediate area. Check all containers of flammable materials in the area to ensure that they are tightly closed.
4. Notify other occupants of the laboratory in advance of lighting a flame.
5. Store flammable materials properly (see Section II.D.2).
6. When volatile flammable materials may be present, use only nonsparking electrical equipment.

I.A.10 COLD TRAPS AND CRYOGENIC HAZARDS

The primary hazard of cryogenic materials is their extreme coldness. They, and surfaces they cool, can cause severe burns if allowed to contact the skin. Gloves and a face shield may be needed when preparing or using some cold baths (see Sections I.F.1 and 2).

Neither liquid nitrogen nor liquid air should be used to cool a flammable mixture in the presence of air because oxygen can condense from the air, which leads to an explosion hazard. Appropriate dry gloves should be used when handling dry ice (see Section II.B.2), which should be added slowly to the liquid portion of the cooling bath to avoid foaming over. Workers should avoid lowering their head into a dry ice chest: carbon dioxide is heavier than air, and suffocation can result.

I.A.11 SYSTEMS UNDER PRESSURE

Reactions should never be carried out in, nor heat applied to, an apparatus that is a closed system unless it is designed and tested to withstand pressure (see Section I.D.2). Pressurized apparatus should have an appropriate relief device. If the reaction cannot be opened directly to the air, an inert gas purge and bubbler system should be used to avoid pressure buildup.

I.A.12 WASTE DISPOSAL PROCEDURES

Laboratory management has the responsibility for establishing waste disposal procedures for routine and emergency situations (see Chapters II.E and G) and communicating these to laboratory workers. Workers should follow these procedures with care, to avoid any safety hazards or damage to the environment.

I.A.13 WARNING SIGNS AND LABELS

Laboratory areas that have special or unusual hazards should be posted with warning signs. Standard signs and symbols have been established for a number of special situations, such as radioactivity hazards, biological hazards, fire hazards, and laser operations. Other signs should be posted to show the locations of safety showers, eyewash stations, exits, and fire extinguishers. Extinguishers should be labeled to show the type of fire for which they are intended (see Section I.F.1). Waste containers should be labeled for the type of waste that can be safely deposited.

The safety- and hazard-sign systems in the laboratory should enable a person unfamiliar with the usual routine of the laboratory to escape in an emergency (or help combat it, if appropriate) (see Section I.F.5).

When possible, labels on containers of chemicals should contain information on the hazards associated with use of the chemical (see Chapters I.B and C). Unlabeled bottles of chemicals should not be opened; such materials should be disposed of promptly and will require special handling procedures (see Chapters II.E and G).

I.A.14 UNATTENDED OPERATIONS

Frequently, laboratory operations are carried out continuously or overnight. It is essential to plan for interruptions in utility services such as electricity, water, and inert gas. Operations should be designed to be safe, and plans should be made to avoid hazards in case of failure. Wherever possible, arrangements for routine inspection of the operation should be made and, in all cases, the laboratory lights should be left on and an appropriate sign should be placed on the door.

One particular hazard frequently encountered is failure of cooling water supplies. A variety of commercial or homemade devices can be used that (a) automatically regulate water pressure to avoid surges that might rupture the water lines or (b) monitor the water flow so that its failure will automatically turn off electrical connections and water supply valves.

I.A.15 WORKING ALONE

Generally, it is prudent to avoid working in a laboratory building alone. Under normal working conditions, arrangements should be made between individuals working in separate laboratories outside of working hours to crosscheck periodically. Alternatively, security guards may be asked to check on the laboratory worker. Experiments known to be hazardous should not be undertaken by a worker who is alone in a laboratory.

Under unusual conditions, special rules may be necessary. The supervisor of the laboratory has the responsibility for determining whether the work requires special safety precautions, such as having two persons in the same room during a particular operation.

I.A.16 ACCIDENT REPORTING

Emergency telephone numbers to be called in the event of fire, accident, flood, or hazardous chemical spill should be posted prominently in each laboratory. In addition, the numbers of the laboratory workers and their supervisors should be posted. These persons should be notified immediately in the event of an accident or emergency.

Every laboratory should have an internal accident-reporting system to help discover and correct unexpected hazards (see Section I.F.5). This system should include provisions for investigating the causes of injury and any potentially serious incident that does not result in injury. The goal of such investigations should be to make recommendations to improve safety, not to assign blame for an incident. Relevant federal, state, and local regulations may require particular reporting procedures for accidents or injuries.

I.A.17 EVERYDAY HAZARDS

Finally, laboratory workers should remember that injuries can and do occur outside the laboratory or other work area. It is important that safety be practiced in offices, stairways, corridors, and other places. Here, safety is largely a matter of common sense, but a constant safety awareness of everyday hazards is vital.

I.A.18 SELECTED BIBLIOGRAPHY

1. Committee on Chemical Safety. *Safety in Academic Chemistry Laboratories*, 3rd ed.; American Chemical Society: Washington, D.C., 1979.

2. Green, M.E.; Turk, A. Safety in Working with Chemicals; McMillan: New York, N.Y., 1978.

3. Muir, G.D., Ed. *Hazards in the Chemical Laboratory*, 2nd ed.; Chemical Society: London, 1972.

4. Steere, N.V., Ed. *CRC Handbook of Laboratory Safety*, 2nd ed.; CRC Press: West Palm Beach, Fla., 1971.

(See also Sections I.E.3 and I.H.7)

I.B

Procedures for Working with Substances that Pose Hazards Because of Acute Toxicity, Chronic Toxicity, or Corrosiveness

Many of the chemicals encountered in the laboratory are known to be toxic or corrosive or both. New and untested substances that may be hazardous are also frequently encountered. Thus, it is essential that all laboratory workers understand the types of toxicity, know the routes of exposure, and recognize the major classes of toxic and corrosive chemicals.

When considering possible toxicity hazards while planning an experiment, it is important to recognize that the combination of the toxic effects of two substances may be significantly greater than the toxic effect of either substance alone. Because most chemical reactions are likely to contain mixtures of substances whose combined toxicities have never been evaluated, it is prudent to assume that mixtures of different substances (i.e., chemical reaction mixtures) will be more toxic than the most toxic ingredient contained in the mixture. Furthermore, chemical reactions involving two or more substances may form reaction products that are significantly more toxic than the starting reactants. This possibility of generating toxic reaction products may not be anticipated by the laboratory worker in cases where the reactants are mixed unintentionally. For example, successive treatments of a surface or an object with aqueous ammonia and then with sodium hypochlorite or other positive-halogen reagents may generate hydrazine, a substance that poses both acute and chronic toxicity hazards. Similarly, inadvertent mixing of formaldehyde (a common tissue fixative) and hydrogen chloride could result in the generation of bis(chloromethyl)ether, a potent human carcinogen.

Exposure to hazardous chemicals can be prevented or minimized by following the procedures detailed elsewhere in this report, especially Chapters I.A, F, and H. Only the most pertinent measures are stressed in this chapter.

I.B.1 ROUTES OF EXPOSURE

Exposure to chemicals may occur by the following routes:

1. inhalation,
2. ingestion,
3. contact with skin and eyes, or
4. injection.

INHALATION

Inhalation of toxic vapors, mists, gases, or dusts can produce poisoning by absorption through the mucous membrane of the mouth, throat, and lungs and can seriously damage these tissues by local action. Inhaled gases or vapors may pass rapidly into the capillaries of the lungs and be carried into the circulatory system. This absorption can be extremely rapid. The rate will vary with the concentration of the toxic substance, its solubility in tissue fluids, the depth of respiration, and the amount of blood circulation, which means that it will be much higher when the person is active than when he or she is at rest.

The degree of injury resulting from exposure to toxic vapors, mists, gases, and dusts depends on the toxicity of the material and its solubility in tissue fluids, as well as on its concentration and the duration of exposure. Chemical activity and the time of response after exposure are not necessarily a measure of the degree of toxicity. Several chemicals (e.g., mercury and its derivatives) and some of the common solvents (benzene) are cumulative poisons that can produce body damage through exposure to small concentrations over a long period of time.

The American Conference of Governmental Industrial Hygienists (ACGIH) produces annual lists of Threshold Limit Values (TLVs®) and Short Term Exposure Limits (STELs) for common chemicals used in laboratories. These values are guides, not legal standards, and are defined as follows:

TLV® Time-weighted average concentration for a normal 8-hour workday to which nearly all workers may be repeatedly exposed without adverse effect.

STEL Maximum concentration to which workers can be exposed for periods up to 15 min. Such exposures should be limited to no more than four per day with periods of at least 60 min each between exposures; the total time-weighted exposure per day should not exceed the TLV value.

The 1980 TLVs are appended to this report as Appendix B.

Most of the 1968 TLVs were adopted by OSHA in 1972 as legal Permissible Exposure Levels (PELs). The basis for selection of the TLVs appears to be more secure than the justification for the STELs. The TLVs provide a useful estimate of how much ventilation may be needed in laboratories where the occupants typically spend most of their working time.

However, because of the many factors influencing toxicity, each situation should be evaluated individually and the TLVs used as guidelines rather than as fine lines between safe and dangerous concentrations.

The best way to avoid exposure to toxic vapors, mists, gases, and dusts is to prevent the escape of such materials into the working atmosphere and to ensure adequate ventilation by the use of exhaust hoods and other local ventilation (see Section I.H.2). Chemicals of unknown toxicity should not be smelled.

Ingestion

Many of the chemicals used in the laboratory are extremely dangerous if they enter the mouth and are swallowed (see Section I.F.7).

The relative acute toxicity of a chemical can be evaluated by determining its LD_{50}, which is defined as the quantity of material that, when ingested or applied to the skin in a single dose, will cause the death of 50% of the test animals. It is expressed in grams or milligrams per kilogram of body weight (see Section I.E.2 for examples of LD_{50} values). In addition, many chemicals may damage the tissues of the mouth, nose, throat, lungs, and gastrointestinal tract and produce systemic poisoning if absorbed through the tissues.

To prevent entry of toxic chemicals into the mouth, laboratory workers should wash their hands before eating, smoking, or applying cosmetics; immediately after use of any toxic substance; and before leaving the laboratory. Food and drink should not be stored or consumed in areas where chemicals are being used nor should cigarettes, cigars, and pipes be used in such areas (see Sections I.A.2 and 3); chemicals should not be tasted; and pipetting and siphoning of liquids should never be done by mouth.

Contact with Skin and Eyes

Contact with the skin is a frequent mode of chemical injury. A common result of skin contact is a localized irritation (see Section I.B.4), but an appreciable number of materials are absorbed through the skin with sufficient rapidity to produce systemic poisoning. The main portals of entry for chemicals through the skin are the hair follicles, sebaceous glands, sweat glands, and cuts or abrasions of the outer layers of the skin. The follicles and glands are abundantly supplied with blood vessels, which facilitates the absorption of chemicals into the body.

Contact of chemicals with the eyes is of particular concern because these organs are so sensitive to irritants. Few substances are innocuous in contact with the eyes; most are painful and irritating, and a considerable number are capable of causing burns and loss of vision. Alkaline materials, phenols, and strong acids are particularly corrosive and can cause permanent loss of vision. Also, eyes are very vascular and provide for rapid absorption of many chemicals.

Skin and eye contact with chemicals should be avoided by use of appropriate protective equipment (see Sections I.F.1-3). All persons in the laboratory should wear safety glasses. Face shields, safety goggles, shields, and similar devices provide better protection for the eyes. Protection against skin contact may be obtained by use of gloves, laboratory coats, tongs, and other protective devices. Spills should be cleaned up promptly (see Section I.F.5).

In the event of skin contact, the affected areas should be flushed with water and medical attention should be sought if symptoms persist; in the event of eye contact, the eye(s) should be flushed with water for 15 min and medical attention should be sought whether or not symptoms persist (see Section I.F.6).

Injection

Exposure to toxic chemicals by injection seldom occurs in the chemical laboratory. However, it can inadvertently occur through mechanical injury from glass or metal contaminated with chemicals or when chemicals are handled in syringes.

I.B.2 ACUTE AND CHRONIC TOXICITY

The toxicity of a material is due to its ability to damage or interfere with the metabolism of living tissue. An acutely toxic substance can cause damage as the result of a single or short-duration exposure. A chronically toxic substance causes damage after repeated or long-duration exposure or that becomes evident only after a long latency period. Hydrogen cyanide, hydrogen sulfide, and nitrogen dioxide are examples of acute poisons. Chronic poisons include all carcinogens and many metals and their compounds (such as mercury and lead and their derivatives). Chronic toxins are particularly insidious because of their long latency periods. The cumulative effects of low exposures may not be apparent for many years. Some chemicals, e.g., vinyl chloride, can be either acutely or chronically toxic, depending on the degree of exposure. All new and untested chemicals should be regarded as toxic until proven otherwise.

I.B.3 EMBRYOTOXINS

Embryotoxins are substances that act during pregnancy to cause adverse effects on the fetus. These effects include embryolethality (death of the fertilized egg, the embryo, or the fetus), malformations (teratologic effects), retarded growth, and postnatal functional deficits.

A few substances have been demonstrated to be embryotoxic in humans. These include organomercurials, lead compounds, and the formerly used sedative, thalidomide. Maternal alcoholism is probably the leading known cause of embryotoxic effects in humans, but the exposure to ethanol encountered in laboratories is unlikely to be embryotoxic. Many substances, some common (e.g., sodium chloride), have been shown to be embryotoxic to animals at some exposure level, but usually this is at a considerably higher level than is met in the course of normal laboratory work. However, some substances do require special controls because of embryotoxic properties. One example is formamide; women of child-bearing potential should handle this substance only in a hood and should take precautions to avoid skin contact with the liquid because of the ease with which it passes through the skin.

Because the period of greatest susceptibility to embryotoxins is the first 8-12 weeks of pregnancy, which includes a period when a woman may not know she is pregnant, women of child-bearing potential should take care to avoid skin contact with all chemicals. The following procedures are recommended to be followed routinely by women of child-bearing potential in working with chemicals requiring special control because of embryotoxic properties:

1. Each use should be reviewed for particular hazards by the research supervisor, who will decide whether special procedures are warranted or whether warning signs should be posted. Consultation with appropriate

safety personnel may be desirable. In cases of continued use of a known embryotoxin, the operation should be reviewed annually or whenever a change in procedures is made.

2. Embryotoxins requiring special control should be stored in an adequately ventilated area. The container should be labelled in a clear manner such as the following: EMBRYOTOXIN: READ SPECIFIC PROCEDURES FOR USE. If the storage container is breakable, it should be kept in an impermeable, unbreakable secondary container having sufficient capacity to retain the material should the primary container accidentally break.

3. Women of child-bearing potential should take adequate precautions to guard against spills and splashes. Operations should be carried out using impermeable containers and in adequately ventilated areas (see Chapter I.H). Appropriate safety apparel (see Sections I.F.1-3), especially gloves, should be worn. All hoods, glove boxes, or other essential engineering controls should be known to be operating at required efficiency before work is started (see Section I.H.2).

4. Supervisors should be notified of all incidents of exposure or spills of embryotoxins requiring special control. A qualified physician should be consulted about any exposures of women of child-bearing potential above the acceptable level, i.e., any skin contact or any inhalation.

I.B.4 ALLERGENS

A wide variety of substances can produce skin and lung hypersensitivity. Examples include such common substances as diazomethane, chromium, nickel, bichromates, formaldehyde, isocyanates, and certain phenols. Because of this variety and because of the varying response of individuals, suitable gloves should be used whenever hand contact with products of unknown activity is probable (see Section I.F.2).

I.B.5 CORROSIVE CHEMICALS

The major classes of corrosive chemicals are strong acids and bases, dehydrating agents, and oxidizing agents. Some chemicals, e.g., sulfuric acid, belong to more than one class. Inhalation of vapors or mists of these substances can cause severe bronchial irritation. These chemicals erode the skin and the respiratory epithelium and are particularly damaging to the eyes.

Strong Acids

All concentrated strong acids can damage the skin and eyes. Exposed areas should be flushed promptly with water. Nitric, chromic, and hydrofluoric acids are especially damaging because of the types of burns they inflict. Hydrofluoric acid, which produces slow-healing, painful burns, should be used only after thorough familiarization with recommended handling procedures (see Section III.F.9).

Strong Bases

The common strong bases are potassium hydroxide, sodium hydroxide, and ammonia. Ammonia is a severe bronchial irritant and should always be used in a well-ventilated area. The metal hydroxides are extremely damaging to the eyes. Should exposure occur, the affected areas should be washed at once with copious quantities of water and an opthalmologist should evaluate the need for further treatment.

Dehydrating Agents

The strong dehydrating agents include concentrated sulfuric acid, sodium hydroxide, phosphorus pentoxide, and calcium oxide. Because much heat is evolved on mixing these substances with water, mixing should always be done by adding the agent to water to avoid violent reaction and spattering. Because of their affinity for water, these substances cause severe burns on contact with the skin. Affected areas should be washed promptly with large volumes of water.

Oxidizing Agents

In addition to their corrosive properties, powerful oxidizing agents such as perchloric and chromic acids (sometimes used as cleaning solution), present fire and explosion hazards on contact with organic compounds and other oxidizable substances. The hazards associated with the use of perchloric acid are especially severe (see also Section I.C.3); it should be handled only after thorough familiarization with recommended procedures. Strong oxidizing agents should be stored and used in glass or other inert containers (preferably unbreakable), and corks and rubber stoppers should not be used. Reaction vessels containing significant quantities of these reagents should be heated by using fiberglass mantles or sand baths rather than oil baths.

I.B.6 TYPES OF HANDLING PROCEDURES

Recommendations for handling procedures for chemicals begin with the admonition that, even for substances of no known significant hazard, it is prudent to observe good laboratory practice, minimizing exposure by working in an exhaust hood and wearing eye and hand protection and a laboratory coat or apron. For the case of substances that present special hazards, we call attention to the following procedures for minimizing risk.

For *toxic substances*, the procedures fall into three groups:

1. OSHA has published detailed procedures (29 CFR 1910; compare also 45 Fed. Reg. 5002-5296) for working with substances they have classified as carcinogens. These standards are more stringent than the guidelines given in this report as Procedure A (see Section I.B.8). Anyone contemplating work with materials on this list should consult the regulations to be advised of the necessary approvals, training, working conditions, monitoring, record-keeping, and medical surveillance. In addition, if a worker anticipates that an OSHA-regulated carcinogen might be a product or an impurity, the regulations should be consulted; e.g., *N*-phenyl-1,4-benzenediamine (semidine) can contain substantial benzidine impurity.

The OSHA list as of the date of this writing includes the following:

2-Acetylaminofluorene
Acrylonitrile
4-Aminobiphenyl
Asbestos
Benzidine
Bis(chloromethyl)ether
3,3′-Dichlorobenzidine (and its salts)
4-Dimethylaminoazobenzene
Ethylenimine
Inorganic arsenic
Methyl chloromethyl ether
4,4′-Methylene-bis(2-chloroaniline)
α-Naphthylamine
β-Naphthylamine
4-Nitrobiphenyl
N-Nitrosodimethylamine
β-Propiolactone
Vinyl chloride

[In addition, the International Agency for Research on Cancer has published a list of 54 chemicals they consider to pose a carcinogenic risk (*Chemicals and Industrial Processes Associated with Cancer in Humans*; International Agency for Research on Cancer: Lyon; Sept. 1979, Monographs Supplement 1.), and the Environmental Protection Agency has published a preliminary list of about 80 substances to be subject to labeling requirements under the Toxic Substances Act of 1976 (45 Fed. Reg. 42,854).]

2. Procedure A (see Section I.B.8) should be followed in laboratory operations using those substances believed to be moderately to highly toxic, even when used in small amounts. A substance that has caused cancer in humans or has shown high carcinogenic potency in test animals (but for which a regulatory standard has not been issued by OSHA) will generally require the use of Procedure A. However, before choosing Procedure A, other factors, such as the physical form and volatility of the substance, the kind and duration of exposure, and the amount to be used should also be considered.

A substance is deemed to have moderate to high carcinogenic potency in test animals if it causes statistically significant tumor incidence (a) after inhalation exposure of 6-7 hours per day, 5 days per week, for a significant portion of a lifetime to dosages of less than 10 mg/m^3, or (b) after repeated skin application of less than 300 (mg/kg of body weight) per week, or (c) after oral dosages of less than 50 (mg/kg of body weight) per day.

3. Procedure B (see Section I.B.7) should be followed in laboratory operations using substances for which infrequent, small quantities do not constitute a significant carcinogenic hazard but which can be dangerous to those exposed to high concentrations or repeated small doses. A substance that is not known to cause cancer in humans, but which has shown statistically significant, but low, carcinogenic potency in animals, generally should be handled according to Procedure B.

Many of the general recommendations for safe practices in the laboratory are especially applicable to the handling of corrosive substances. It is also important that attention be given to the use of protective apparel and safety equipment (see Chapter I.F). In addition, the storage, disposal, and cleanup of corrosive substances requires special care (see also Chapters II.B, D, and E). Bottles of corrosive liquids should be stored in acid containers or in polyethylene or lead trays or containers large enough to contain the contents of the bottles; most major suppliers will provide acids in plastic-coated glass bottles, which are much less likely to break than ordinary bottles. To ensure that mutually reactive chemicals cannot accidentally contact one another, such substances should not be stored in the same trays unless they are in unbreakable, corrosion-resistant secondary containers. In disposal of corrosive substances, care must be taken not to mix them with other potentially reactive wastes. In most cases, spills involving these substances should be contained, carefully diluted with water, and then neutralized (see Section II.E.6).

I.B.7 GENERAL PROCEDURES AND PRECAUTIONS FOR WORKING WITH SUBSTANCES OF MODERATE CHRONIC OR HIGH ACUTE TOXICITY (PROCEDURE B)

Before beginning a laboratory operation, each worker is strongly advised to consult one of the standard compilations (see Section I.E.4) that list toxic properties of known substances and learn what is known about the substances that will be used. The precautions and procedures described below (termed Procedure B in this report) should be followed if any of the substances to be used in significant quantities is known to be moderately or highly toxic. (If any of the substances being used is known to be highly toxic, it is desirable that there be two people present in the area at all times.) These procedures should also be followed if the toxicological properties of any of the substances being used or prepared are unknown. If any of the substances to be used or prepared are known to have high chronic toxicity (e.g., compounds of heavy metals and strong carcinogens), then the precautions and procedures described below should be supplemented with additional precautions (termed Procedure A in this report; see Section I.B.8) to aid in containing and, ultimately, destroying substances having high chronic toxicity.

The overall objective of Procedure B is to minimize exposure of the laboratory worker to toxic substances, by any route of exposure, by taking all reasonable precautions. Thus, the general precautions outlined in other chapters (especially I.A, F, and H) should normally be followed whenever a toxic substance is being transferred from one container to another or is being subjected to some chemical or physical manipulation. The following three precautions should always be followed:

1. Protect the hands and forearms by wearing either gloves and a laboratory coat or suitable long gloves (gauntlets) (see Section I.F.2) to avoid contact of toxic material with the skin.

2. Procedures involving volatile toxic substances and those involving solid or liquid toxic substances that may result in the generation of aerosols should be conducted in a hood or other suitable containment device (see Sections I.H.2, 4, and 5). (The hood should have been

evaluated previously to establish that it is providing adequate ventilation and has an average face velocity of not less than 60 linear ft/min.)

3. After working with toxic materials, wash the hands and arms immediately. Never eat, drink, smoke, chew gum, apply cosmetics, take medicine, or store food in areas where toxic substances are being used.

These standard precautions will provide laboratory workers with good protection from most toxic substances. In addition, records that include amounts of material used and names of workers involved should be kept as part of the laboratory notebook record of the experiment. To minimize hazards from accidental breakage of apparatus or spills of toxic substances in the hood, containers of such substances should be stored in pans or trays made of polyethylene or other chemically resistant material and apparatus should be mounted above trays of the same type of material. Alternatively, the working surface of the hood can be fitted with a removable liner of adsorbent plastic-backed paper. Such procedures will contain spilled toxic substances in a pan, tray, or adsorbent liner and greatly simplify subsequent cleanup and disposal. Vapors that are discharged from the apparatus should be trapped or condensed to avoid adding substantial amounts of toxic vapor to the hood exhaust air. Areas where toxic substances are being used and stored should have restricted access, and special warning signs should be posted if a special toxicity hazard exists.

The general waste-disposal procedures described in this report (see Chapter II.E) should be followed. However, certain additional precautions should be observed when waste materials are known to contain substances of moderate or high toxicity. Volatile toxic substances should never be disposed of by evaporation in the hood. If practical, waste materials and waste solvents containing toxic substances should be decontaminated chemically by some procedure that can reasonably be expected to convert essentially all of the toxic substance to nontoxic substances. If chemical decontamination is not feasible, the waste materials and solvents containing toxic substances should be stored in closed, impervious containers so that personnel handling the containers will not be exposed to their contents. In general, liquid residues should be contained in glass or polyethylene bottles half-filled with vermiculite. In some institutions, nonreactive liquids are incinerated in a solvent burner. In such cases, it may be appropriate to add used motor oil to reduce flammability and vermiculite or other solids should not be added. All containers of toxic wastes should be suitably labeled to indicate the contents (chemicals and approximate amounts) and the type of toxicity hazard that contact may pose. For example, containers of wastes from experiments involving appreciable amounts of weak or moderate carcinogens should carry the warning: CANCER-SUSPECT AGENT. All wastes and residues that have not been chemically decontaminated in the exhaust hood where the experiment was carried out should be disposed of in a safe manner that ensures that personnel are not exposed to the material (typically by incineration; see Section II.G.1).

The laboratory worker should be prepared for possible accidents or spills involving toxic substances. If a toxic substance contacts the skin, the area should be washed well with water or a safety shower should be used. If there is a major spill outside the hood, the room or appropriate area should be evacuated and necessary measures (see Section II.E.6) to prevent exposure of other workers should be taken. Spills should be cleaned up by personnel wearing suitable personal protective apparel (see Section I.F.3). If a spill of a toxicologically significant quantity of toxic material occurs outside the hood, a supplied-air full-face respirator should be worn. Contaminated clothing and shoes should be thoroughly decontaminated or incinerated.

EXAMPLES

Diisopropyl Fluorophosphate

Diisopropyl fluorophosphate (DFP) is a valuable reagent in biochemical laboratories because it is an effective in vitro inhibitor of serine-type proteolytic enzymes useful in curtailing destruction of proteins.

Physical Properties

DFP is a viscous liquid having a vapor pressure of 0.579 at 20°C. It is soluble in water (1.54% v, w/w), but decomposes in this medium at a finite rate to form hydrofluoric acid.

Toxicity

DFP, as well as the organophosphorus insecticides, reacts with acetylcholinesterase to form covalent derivatives, which leads to inactivation of the enzyme and the accumulation of acetylcholine. Fortunately, the toxic effects respond to treatment if the proper antidotes are immediately available.

DFP is absorbed by inhalation, ingestion, or topical contact. It causes pin-point pupils in those affected, as well as severe lacrymation and rhinitis. Contact with skin can lead to localized fasciculations of the muscles nearby. Other symptoms of exposure include weakness, wheezing, and tachycardia. Severe intoxication is evidenced by ataxia, confusion, convulsions, and respiratory paralysis leading to death.

Handling Procedures

DFP should be obtained in ampules containing the minimum amount necessary for use; the reagent is not expensive, and excess material should be neutralized and disposed of in a safe manner. Ampules containing 0.1, 0.4, and 1.0 g are available. The ampules should be opened in a hood using great care to avoid splatter or in a glove box equipped with a HEPA filter. Protective apparel should include disposable gloves and a face shield.

A large container of 2 *N* NaOH should be available wherever DFP is being used so that the entire ampule, if necessary, can be discarded into it; decomposition of DFP is relatively rapid in basic solution. Once the ampule has been opened, the required amount of DFP should be withdrawn with a disposable syringe and used immediately. The opened ampule and gloves and other items that have come in contact with the DFP should be washed with base before disposal. Any spills that occur should be neutralized with excess base before they are cleaned up.

Solutions of DFP should be handled only under conditions that provide adequate ventilation. Most cold rooms, for example, are not adequately ventilated. All solutions of DFP should be kept in tightly stoppered containers. After such solutions have served their purpose, they should be made alkaline to ensure complete decomposition of any remaining DFP before disposal.

Emergency Treatment

Treatment consists of immediate administration of atropine followed by a specific antidote, pyridine-2-aldoxime (2-PAM; pralidoxime chloride; commercially available as Protopam Chloride). Atropine is used to block certain receptors upon which acetylcholine acts while pyridine-2-aldoxime reactivates the covalently modified acetylcholinesterase. Both drugs should be on hand in any laboratory in which DFP is used, and medical assistance must be readily available.

The keys to safety in handling DFP are an appreciation of its toxicity, the presence of adequate ventilation, a solution of base in case of spillage and for discarding all implements in contact with the reagent, and the presence at the site of the two antidotes.

Hydrofluoric Acid

All forms— dilute or concentrated solutions or the vapor— of hydrofluoric acid (HF) cause severe burns. Inhalation of anhydrous HF or HF mists or vapors can cause severe respiratory tract irritation that may be fatal.

Physical Properties

Anhydrous HF (formula weight 20.01) is a clear, colorless liquid that boils at 19.5°C. Because of its low boiling point and high vapor pressure, anhydrous HF must be stored in pressure containers. A 70% aqueous solution is a common form of HF. Hydrofluoric acid is miscible with water in all proportions and forms an azeotrope (38.3% HF) that boils at 112°C.

Toxicity

Anhydrous or concentrated aqueous HF causes immediate and serious burns to any part of the body. Dilute solutions and gaseous HF are also harmful, although several hours may pass before the HF penetrates the skin sufficiently to cause redness or a burning sensation.

Wearing clothing (including leather shoes and gloves) that has absorbed small amounts of HF can result in serious delayed effects such as painful, slow-healing skin ulcers.

Hazards from Fire or Explosion

Hydrofluoric acid is nonflammable. It is difficult to contain because it attacks glass, concrete, and some metals— especially cast iron and alloys that contain silica. It also attacks such organic materials as leather, natural rubber, and wood. Because aqueous HF can cause formation of hydrogen in metallic containers and piping, which presents a fire and explosion hazard, potential sources of ignition (sparks and flames) should be excluded from areas having equipment containing HF.

Handling Procedures

It is crucial to ensure adequate ventilation by working only in a hood so that safe levels (3 ppm) are not exceeded. All contact of the vapor or the liquid with eyes, skin, respiratory system, or digestive system must be avoided by using protective equipment such as face shields and neoprene or polyvinyl chloride gloves. The protective equipment should be washed after each use to remove any HF on it. Safety showers and eyewash fountains should be nearby. Anyone working with HF should have received prior instruction about its hazards and in proper protective measures and should know the recommended procedure for treatment in the event of exposure (Reinhardt, C. F. *et al.*; Am. Ind. Hyg. Assn. J., 1966, 27, 166.).

Spills and Leaks—The vapors of both anhydrous HF and aqueous 70% HF produce visible fumes if they contact moist air. This characteristic can be useful in detecting leaks but cannot be relied on because of atmospheric variations. Spills of HF must be treated immediately to minimize the dangers of vapor inhalation, body contact, corrosion of equipment, and possible generation of hazardous gases. Spills should be contained and diluted with water. The resulting solution should be neutralized with lime before disposal.

Waste disposal—Waste HF may be added slowly to a larger volume of an agitated solution of slaked lime. This neutralized solution is then added to excess running water before final disposal. Because sodium fluoride is highly soluble and toxic to warm-blooded animals, lime is the preferred neutralizing agent.

Emergency Treatment

Anyone who knows or even suspects that he or she has come into direct contact with HF should immediately flush the exposed area with large quantities of cool water. Exposed clothing should be removed as quickly as possible while flushing. Medical attention should be obtained promptly, even if the injury appears slight. On the way to the physician, the burned area should be immersed in a mixture of ice and water or, if readily available, an iced solution of benzalkonium chloride. (It may be necessary to remove the area being soaked from the solution periodically to relieve discomfort caused by the cold.) If immersion is impractical, a compress made by inserting ice cubes between layers of gauze should be used.

If HF liquid or vapor has contacted the eyes, these organs should be flushed with large quantities of clean water while the eyelids are held apart. This flushing should be continued for 15 min. Medical attention should be obtained promptly.

Anyone who has inhaled HF vapor should be removed *immediately* to an uncontaminated atmosphere and kept warm. Medical help should be obtained promptly. Anyone who has ingested HF should drink a large quantity of water as quickly as possible. Do *not* induce vomiting. Again, medical help should be obtained promptly. After the acid has been thoroughly diluted with water, if medical attention is delayed, the person should be given milk or two fluid ounces of milk of magnesia to drink to sooth the burning effect.

Hydrogen Cyanide

The procedures given below are appropriate for the safe use of hydrogen cyanide (HCN) and related compounds (e.g., cyanogen and cyanogen halides) that may release cyanide ion or cyanogen or generate HCN when acidified (e.g., sodium or potassium cyanide). The use of the term HCN in these procedures is intended to be specific for hydrogen cyanide and to indicate general applicability to related compounds. All users of HCN should study or be instructed in HCN procedures before starting to work with this material.

Physical Properties

See Safety Data Sheet (Section I.E.2).

Toxicity

See Safety Data Sheet (Section I.E.2).

Hazards from Fire, Explosion, or Uncontrolled Polymerization

Because of its low flash point and wide range of explosive mixtures, HCN presents a serious fire and explosion hazard. Another often overlooked hazard is spontaneous polymerization, which results from the attack of cyanide ion on HCN. The reaction is unpredictable, but may occur in pure unstabilized HCN if a base is present. Hence, amines, hydroxides, and cyanide salts that are capable of producing the cyanide ion should not be added to liquid HCN without suitable precautions. Once started, the polymerization is likely to become violent and be accompanied by sharp increases in temperature and pressure. If liquid HCN is heated above 115°C in a sealed vessel, a violent exothermic reaction generally occurs.

Commercial HCN is normally stabilized by the addition of a little phosphoric acid, although sulfuric acid or sulfur dioxide may be used also. Either acid may be removed readily by distillation, but sulfur dioxide is extremely difficult to remove. Distilled HCN constitutes a greater explosion hazard than stabilized materials, so use of the distilled material should be avoided.

Handling Procedures

All work with HCN must be confined to hoods, which should have a minimum face velocity of 60 lfm. Care must be exercised to prevent the contact of either liquid HCN or its vapors with the skin. Neoprene or rubber gloves should be worn at all times when working with HCN. Whenever work with HCN or related compounds is being carried out in a laboratory, there should be at least *two people present in the area at all times.*

Signs warning that HCN is in use should be posted at each entrance to the laboratory area whenever work is being done with HCN. WARNING or NO ADMITTANCE signs should be posted on the doors to fan lofts and roofs whenever cyanides are being used or stored in hoods.

All reaction equipment in which cyanides are used or produced should be placed in or over shallow pans so that spills or leaks will be contained. In the event of spills of HCN or cyanide solutions, the contaminated area should be evacuated promptly and it should be determined immediately whether anyone has been exposed to cyanide vapors or liquid splash. Consideration should be given to the need for evacuating other parts of the building or notifying other occupants that the spill has occurred. In general, it is usually best not to attempt to dilute or absorb such spills if they occur in well-ventilated areas.

Detection—Hydrogen cyanide has a characteristic odor that resembles that of bitter almonds; however, many people cannot smell it in low concentrations, and this method of detection should not be relied on. Vapor-detector tubes sensitive to 1 ppm of HCN are available commercially. The presence of free cyanide ion in aqueous solution may be detected by treating an aliquot of the sample with ferrous sulfate and an excess of sulfuric acid. A precipitate of Prussian blue indicates that free cyanide ion is present.

Sodium cyanide and acids should not be stored or transported together. An open bottle of NaCN can generate HCN in humid air, and HCN may be liberated from spills of sodium cyanide solutions.

Storage and dispensing—Storage of liquid HCN (except in commercial cylinders) in laboratory areas should be prohibited without special permission. If such storage is necessary, it must be in an exhaust hood or a barricaded area that has independent ventilation facilities. Liquid HCN is dispensed from cylinders. Only trained personnel should be permitted to operate the dispensing equipment.

Waste disposal—Hydrogen cyanide and waste solutions containing cyanides must not be emptied into the sewer nor left to evaporate in an exhaust hood. The most effective way to dispose of such material is to dilute it with an equal amount of alcohol and burn it in a solvent incinerator.

Emergency Treatment

The following equipment and procedures are recommended:

Equipment—An HCN first aid kit and an oxygen cylinder equipped with pressure gage and needle valve should be available on any floor of a building on which work with cyanides is in progress. In special cases, first aid kits and oxygen cylinders may be located near the work area but they should not be in the same room. Except when the cylinder is being used or checked, the main cylinder valve should be kept closed and the pressure kept off the gage. A tag should be attached to the oxygen cylinder indicating that it is reserved for emergency HCN first aid.

The HCN first aid kit should contain a box of amyl nitrite pearls, a facepiece and length of rubber tubing for administering oxygen, and a bottle of 1% sodium thiosulfate solution.

Procedures—Anyone who has been exposed to HCN should be removed from the contaminated atmosphere immediately. Any contaminated clothing should be removed and the affected area deluged with water. The person should be kept warm. An amyl nitrite pearl should be held under the person's nose for not more than 15 s/min (excess nitrite will reduce the blood pressure), and oxygen should be administered in the intervals. If the person is not breathing, artificial resuscitation should be begun; when breathing starts, amyl nitrite and oxygen should be administered immediately. If HCN has been ingested, the person should be given one pint of 1% sodium thiosulfate and then soapy water or mustard water to induce vomiting. Such cases must be transported to definitive medical care immediately.

I.B.8 ADDITIONAL PROCEDURES AND PRECAUTIONS FOR WORKING WITH SUBSTANCES OF KNOWN HIGH CHRONIC TOXICITY (PROCEDURE A)

All of the procedures and precautions described above should be followed when working with substances known to have high chronic toxicity (Procedure B). In addition, when such substances are to be used in quantities in excess of a few milligrams to a few grams (depending on the hazard posed by the particular substance), the additional precautions described below (termed Procedure A) should be used. Each laboratory worker's plans for experimental work and for disposing of waste materials should be approved by the laboratory supervisor. Consultation with the departmental safety coordinator may be appropriate to ensure that the toxic material is effectively contained during the experiment and that waste materials are disposed of in a safe manner. Substances in this high-chronic-toxicity category include certain heavy metal compounds (e.g., dimethylmercury and nickel carbonyl) and compounds normally classified as strong carcinogens. Examples of compounds frequently considered to be strong carcinogens include the following: benzo[*a*]pyrene; 3-methylcholanthrene; 7,12-dimethylbenz[*a*]anthracene; dimethylcarbamoyl chloride; hexamethylphosphoramide; 2-nitronaphthalene; propane sultone; many *N*-nitrosamines; many *N*-nitrosamides; bis(chloromethyl)ether; aflatoxin B_1; and 2-acetylaminofluorene.

An accurate record of the amounts of such substances being stored and of the amounts used, dates of use, and names of users should be maintained. It may be appropriate to keep such records as part of the record of experimental work in the laboratory workers' research notebooks, but it must be understood that the research supervisor is responsible for ensuring that accurate records are kept. Any volatile substances having high chronic toxicity should be stored in a ventilated storage area in a secondary tray or container having sufficient capacity to contain the material should the primary container accidentally break. All containers of substances in this category should have labels that identify the contents and include a warning such as the following: WARNING! HIGH CHRONIC TOXICITY or CANCER-SUSPECT AGENT. Storage areas for substances in this category should have limited access, and special signs should be posted if a special toxicity hazard exists. Any area used for storage of substances of high chronic toxicity should be maintained under negative pressure with respect to surrounding areas.

All experiments with and transfers of such substances or mixtures containing such substances should be done in a controlled area (see Chapter I.H). (NOTE: A *controlled area* as defined in this report is a laboratory, a portion of a laboratory, or a facility such as an exhaust hood or a glove box that is designated for the use of highly toxic substances. Its use need not be restricted to the handling of toxic substances if all personnel who have access to the controlled area are aware of the nature of the substances being used and the precautions that are necessary.) When a negative-pressure glove box in which work is done through attached gloves is used, the ventilation rate in the glove box should be at least two volume changes per hour, the pressure should be at least 0.5 in. of water lower than that of the external environment, and the exit gases should be passed through a trap or HEPA filter. Positive-pressure glove boxes are normally used to provide an inert anhydrous atmosphere. If these glove boxes are used with highly toxic compounds, then the box should be thoroughly checked for leaks before each use and the exit gases should be passed through a suitable trap or filter. Laboratory vacuum pumps used with substances having high chronic toxicity should be protected by high-efficiency scrubbers or HEPA filters and vented into an exhaust hood. Motor-driven vacuum pumps are recommended because they are easy to decontaminate (Note: decontamination of a vacuum pump should be carried out in an exhaust hood). Controlled areas should be clearly marked with a conspicuous sign such as the following: WARNING: TOXIC SUBSTANCE IN USE or CANCER-SUSPECT AGENT: AUTHORIZED PERSONNEL ONLY. Only authorized and instructed personnel should be allowed to work in or have access to controlled areas.

Proper gloves (see Section I.F.3 and Table 3) should be worn when transferring or otherwise handling substances or solutions of substances having high chronic toxicity. In some cases, the laboratory worker or the research supervisor may deem it advisable to use other protective apparel (see Chapter I.F), such as an apron of reduced permeability covered by a disposable coat. Extreme precautions such as these might be taken, for example, when handling large amounts of certain heavy metals and their derivatives or compounds known to be potent carcinogens. Surfaces on which high-chronic-toxicity substances are handled should be protected from contamination by using chemically resistant trays or pans that can be decontaminated after the experiment or by using dry, absorbent, plastic-backed paper that can be disposed of after use (see Chapter II.E).

On leaving a controlled area, laboratory workers should remove any protective apparel that has been used and thoroughly wash hands, forearms, face, and neck. If disposable apparel or absorbent paper liners have been used, these items should be placed in a closed and impervious container that should then be labeled in some manner such as the following: CAUTION: CONTENTS CONTAMINATED WITH SUBSTANCES OF HIGH CHRONIC TOXICITY. Nondisposable protective apparel should be thoroughly washed, and containers of disposable apparel and paper liners should be incinerated.

Wastes and other contaminated materials from an experiment involving substances of high chronic toxicity should be collected together with the washings from flasks and such and either decontaminated chemically or placed in closed, suitably labeled containers for incineration away from the controlled area. If chemical decontamination is to be used, a method should be chosen that can reasonably be expected to convert essentially all of the toxic materials into nontoxic materials. For example, residues and wastes from experiments in which β-propiolactone, bis(chloromethyl)-

ether, or methyl chloromethyl ether have been used should be treated for 10 min with concentrated aqueous ammonia.

In the event that chemical decontamination is not feasible, wastes and residues should be placed in an impervious container that should be closed and labeled in some manner such as the following: CAUTION: COMPOUNDS OF HIGH CHRONIC TOXICITY or CAUTION: CANCER-SUSPECT AGENT. Transfer of contaminated wastes from the controlled area to the incinerator should be done under the supervision of authorized personnel and in such a manner as to prevent spill or loss. In general, liquid wastes containing such compounds should be placed in glass or polyethylene (usually preferable) bottles half filled with vermiculite and these should be transported in plastic or metal pails of sufficient capacity to contain the material in case of accidental breakage of the primary container.

Normal laboratory work should not be resumed in a space that has been used as a controlled area until it has been adequately decontaminated. Work surfaces should be thoroughly washed and rinsed. If experiments have involved the use of finely divided solid materials, dry sweeping should not be done. In such cases, surfaces should be cleaned by wet mopping or by use of a vacuum cleaner equipped with a high efficiency particulate air (HEPA) filter. All equipment (e.g., glassware, vacuum traps, and containers) that is known or suspected to have been in contact with substances of high chronic toxicity should be washed and rinsed before they are removed from the controlled area.

In the event of continued experimentation with a substance of high chronic toxicity (i.e., if a worker regularly uses toxicologically significant quantities of such a substance three times a week), a qualified physician should be consulted to determine whether it is advisable to establish a regular schedule of medical surveillance or biological monitoring.

In addition to those substances, procedures, and precautions discussed in this section, certain state and federal regulatory agencies have listed substances whose use and disposal is regulated. These lists of regulated substances are usually accompanied by specific requirements for use and disposal. The lists of substances and the requirements for use and disposal are changing frequently. Because compliance with these regulations is required by law, every departmental safety coordinator and all research supervisors should know the current lists of regulated substances and the requirements for their use and disposal.

EXAMPLES

BENZO[*a*]PYRENE (3,4-benzpyrene)

Although the information given below applies specifically to benzo[*a*]-pyrene, the general precautions and procedures are also applicable to other carcinogenic polycyclic aromatic hydrocarbons.

Physical and Chemical Properties

Benzo[*a*]pyrene is a pale yellow crystalline solid that melts at 176-177°C and boils at 310-312°C/10 mm. It is readily soluble in benzene, toluene, xylene, and acetone. It is sparingly soluble in ethanol and methanol and almost insoluble (0.005 mg/l at 27°C) in water. Benzo[*a*]pyrene is light labile and is oxidized by chromic acid and by ozone. [See also Safety Data Sheet (Section I.E.2).]

Toxicity

Benzo[*a*]pyrene is a potent carcinogen and has produced tumors in all of the at least nine species for which experimental tests have been reported. Both local (skin and subcutaneous tissues) and systemic (lung and liver) carcinogenic effects have been observed. Tumors have resulted from topical application to the skin, oral administration, and parenteral injection. Benzo[*a*]pyrene is also mutagenic in systems that possess mixed function oxidase systems for metabolism of the hydrocarbon. [See also Safety Data Sheet (Section I.E.2).]

Handling Procedures

All work with benzo[*a*]pyrene in quantities in excess of a few milligrams or capable of resulting in the formation of aerosols should be carried out in a well-ventilated hood or in a glove box equipped with a HEPA filter. All work should be carried out in apparatus that is contained in or mounted above unbreakable pans that will contain any spill, except that, for very small amounts, a disposable mat may be adequate to contain possible spills. All containers should bear a label such as the following: CANCER-SUSPECT AGENT. All personnel who handle benzo[*a*]pyrene should wear plastic or latex gloves and a fully buttoned laboratory coat.

Storage and use—All bottles of benzo[a]pyrene should be stored and transported in unbreakable outer containers.

Cleanup of spills and waste disposal—Disposal of benzo[*a*]pyrene is best carried out by oxidation; this can be accomplished by high-temperature incineration or by the use of strong oxidants such as chromic acid cleaning solution. The latter procedure is especially applicable to small residues on glassware and such. For incineration of liquid wastes, solutions should be neutralized if necessary, filtered to remove solids, and put in a polyethylene container for transport. All equipment should be thoroughly rinsed with solvent to decontaminate it, and this solvent should be added to the wastes to be incinerated. Great care should be exercised to prevent contamination of the outside of the solvent container.

Solid reaction wastes should be incinerated or decomposed by other oxidation procedures. Alternatively, solid reaction wastes can be extracted with solvent that is added to other liquid waste for incineration. Any contaminated rags, paper, and such should be incinerated. Contaminated solid materials should be enclosed in sealed plastic bags that are labeled CANCER-SUSPECT AGENT and with the name and amount of the carcinogen. The bags should be stored in a well-ventilated area until they are incinerated.

MERCURY AND ITS COMPOUNDS

Metallic mercury (Hg) is widely used in laboratory instruments, and mercury compounds are used in many laboratory experiments.

Physical Properties

Mercury (atomic weight 200.61) is a heavy, silvery white, shining metal that is liquid at ordinary room temperatures. It has a melting point of −38.9°C, a boiling point of 356.9°C, and a specific gravity of 13.595 at 4°C. It is insoluble in water. Its vapors are colorless, odorless, and tasteless.

Toxicity

Metallic Hg and mercury compounds can be absorbed into the body by inhalation, ingestion, or contact with the skin. Mercury is a subtle poison, the effects of which are cumulative and not readily reversible. The maximum exposure level (time-weighted average) for mercury compounds is 0.05 mg/m^3; the TLV for organic mercury compounds is 0.001 ppm.

Mercury poisoning from exposure by chronic inhalation produces a variety of symptoms. The characteristic effects are emotional disturbances, unsteadiness, inflammation of the mouth and gums, general fatigue, memory loss, and headaches. Kidney damage may result from poisoning by mercurical salts.

In most cases of exposure by chronic inhalation, the symptoms of poisoning gradually disappear when the source of exposure is removed. However, improvement may be slow and complete recovery may take years. Skin contact with mercury compounds produces irritation and various degrees of corrosion. Soluble mercury salts can be absorbed through the intact skin and produce poisoning.

Handling Procedures

Every effort should be made to prevent spills of metallic Hg because the substance is extremely difficult and time consuming to pick up. Droplets get into cracks and crevices, under table legs, and under and into equipment. If spills are frequent and Hg is added to the general air level, the combined concentration may exceed the allowable limits.

Storage—Containers of Hg should be kept closed and stored in secondary containers in a well-ventilated area. When breakage of instruments or apparatus containing Hg is a possibility, the equipment should be placed in an enameled or plastic tray or pan that can be cleaned easily and is large enough to contain the Hg. Transfers of Hg from one container to another should be carried out in a hood, over a tray or pan to confine any spills.

Cleanup of Spills—Pools and droplets of metallic Hg can be pushed together and then collected by suction by using an aspirator bulb or a vacuum device made from a filtering flask, a rubber stopper, and several pieces of flexible and glass tubing. Alternatively, mercury-spill cleanup kits are available commercially or small drops can be picked up on cellophane tape. A mercury-vapor analyzer should be available for determining the effectiveness of the cleanup operation.

If Hg has been spilled on the floor, the workers involved in cleanup and decontamination activities should wear plastic shoe covers. When the cleanup is complete, the shoe covers should be disposed of and the workers should thoroughly wash their hands, arms, and face several times.

Spilled mercury compounds or solutions can be cleaned up by any method that does not cause excessive airborne contamination or skin contact.

Waste Disposal—Significant quantities of metallic Hg from spills or broken thermometers or other equipment and contaminated Hg from laboratory activities should be collected in thick-walled high-density polyethylene bottles for reclamation.

Rags, sponges, shoe covers, and such used in cleanup activities, and broken thermometers containing small amounts of residual mercury, should be placed in a sealed plastic bag, labeled, and disposed of in a safe manner.

N-NITROSODIETHYLAMINE (DIETHYLNITROSAMINE)

Although the information given below applies specifically to *N*-nitrosodiethylamine [$(CH_3CH_2)_2$ N-N=O], the general precautions and procedures are also applicable to other dialkylnitrosamines. It should be noted that the use of *N*-nitrosodimethylamine (dimethylnitrosamine) is regulated by OSHA when it is present at levels equal to or greater than 1% of a solution or mixture.

Physical and Chemical Properties

N-Nitrosodiethylamine is a volatile yellow liquid that has a boiling point of 177°C and a density (D_4^{20}) of 0.9422. It is readily soluble in many organic liquids and in lipids and to the extent of approximately 10% in water. It is stable at room temperature for several days in aqueous solution at neutral and alkaline pH. It is less stable at strongly acid pH at room temperature, but it is not readily decomposed under the latter conditions. *N*-Nitrosodiethylamine can be oxidized by strong oxidants to the corresponding nitramine and can be reduced by various reducing agents to the hydrazine or amine.

Toxicity

N-Nitrosodiethylamine is strongly hepatotoxic and can cause death from liver insufficiency in experimental animals. It is carcinogenic in at least 10 animal species, including subprimates. The main targets for its carcinogenic activity are the liver, lung, esophagus, trachea, and nasal cavity. Although data are not available on the toxicity of *N*-nitrosodiethylamine in humans, the closely related compound *N*-nitrosodimethylamine has caused extensive liver damage as a consequence of industrial exposure. The toxicity in experimental animals can be observed as a consequence of ingestion, inhalation, or topical application to the skin.

Handling Procedures

All work with *N*-nitrosodiethylamine should be carried out in a well-ventilated hood or in a glove box equipped with a HEPA filter. To the extent possible, all vessels that contain *N*-nitrosodiethylamine should be kept closed. All work should be carried out in apparatus that is contained in or mounted above unbreakable pans that will contain any spill. All containers should bear a label such as the following: CANCER-SUSPECT AGENT. All personnel who handle the material should wear plastic, latex, or neoprene gloves and a fully buttoned laboratory coat.

Storage—All bottles of *N*-nitrosodiethylamine should be stored and transported within an unbreakable outer container; storage should be in a ventilated storage cabinet (or in a hood).

Cleanup of spills and waste disposal—Because *N*-nitrosodiethylamine is chemically stable under usual conditions, disposal is best carried out by incineration. For incineration of liquid wastes, solutions should be neutralized if necessary, filtered to remove solids, and put in closed polyethylene containers for transport. All equipment should be thoroughly rinsed with solvent, which should then be added to the liquid waste for incineration. Great care should be exercised to prevent contamination of the outside of the solvent container. If possible, solid wastes should be incinerated; if this is not possible, solid wastes from reaction mixtures that may contain *N*-nitrosodiethylamine should be extracted and the extracts added to the liquid waste. Similarly, any rags, paper, and such that may be contaminated should be incinerated. Contaminated solid materials should be enclosed in sealed plastic bags that are labeled CANCER-SUSPECT AGENT and with the name and amount of the carcinogen. The bags should be stored in a well-ventilated area until they are incinerated.

Spills of *N*-nitrosodiethylamine can be absorbed by Celite® or a commercial spill absorbant. After the absorbant containing the major share of the nitrosamine has been picked up (avoid dusts; do not sweep), the surface should be thoroughly cleaned with a strong detergent solution. If a major spill occurs outside of a ventilated area, the room should be evacuated and the cleanup operation should be carried out by persons equipped with self-contained respirators. Those involved in this operation should wear rubber gloves, laboratory coats, and plastic aprons or equivalent protective apparel.

I.B.9 SPECIAL PROCEDURES AND PRECAUTIONS FOR WORKING WITH SUBSTANCES OF HIGH CHRONIC TOXICITY IN EXPERIMENTAL ANIMALS

The use of substances of high chronic toxicity in experimental animals can present a special exposure hazard, in particular because of the possibility of the formation of aerosols or dusts that contain the toxic substance. Such dusts and aerosols can become dispersed throughout the laboratory or animal quarters through the animal food, urine, or feces. Accordingly, procedures should be devised that reduce the formation of such aerosols and dusts to the lowest possible level. All procedures should be designed to minimize the possible exposure of personnel.

Administration of the substances by injection or gavage is preferable. However, if it is to be administered in the diet, a relatively closed caging system, either one in which the cages are under negative pressure or one in which there is a horizontal laminar airflow directed toward HEPA filters, should be used. Procedures such as the use of a vacuum cleaner equipped with a HEPA filter or wetting the bedding to reduce dusts should be used for the removal of contaminated bedding or cage matting. All toxic-substance-containing diets should be mixed within closed containers and within a hood.

Workers carrying out such operations should wear plastic or rubber gloves and a fully buttoned laboratory coat or equivalent clothing at all times. If exposure to aerosols cannot be controlled in other ways, a respirator (see Section I.F.4) should be used.

When large-scale studies are being carried out with highly toxic substances, special facilities or rooms having restricted access are preferable. If the caging system for any test animals does not adequately protect the personnel, the use of a jumpsuit or similar clothing and shoe and head coverings should be considered.

(See also *Chemical Carcinogen Hazards in Animal Research Facilities*, Office of Biohazard Safety, National Cancer Institute, March 1979.)

I.C

Procedures for Working with Substances that Pose Hazards Because of Flammability or Explosibility

Flammable substances are among the most common of the hazardous materials found in laboratories. However, the ability to vaporize, ignite, and burn or to explode varies with the specific type or class of substance. Prevention of fires and explosions requires knowledge of the flammability characteristics (limits of flammability, ignition requirements, and burning rates) of combustible materials likely to be encountered under various conditions of use (or misuse) and of the appropriate procedures to use in handling such substances.

I.C.1 FLAMMABILITY AND EXPLOSIBILITY OF MIXTURES OF AIR WITH GASES, LIQUIDS, AND DUSTS

PROPERTIES OF FLAMMABLE SUBSTANCES

Flammable substances are those that readily catch fire and burn in air. A flammable liquid does not itself burn; it is the vapors from the liquid that burn. The rate at which different liquids produce flammable vapors depends on their vapor pressure, which increases with temperature. The degree of fire hazard depends also on the ability to form combustible or explosive mixtures with air, the ease of ignition of these mixtures, and the relative densities of the liquid with respect to water and of the gas with respect to air. These concepts can be evaluated and compared in terms of a number of properties.

Flash Point

An open beaker of diethyl ether set on the laboratory bench next to a Bunsen burner will ignite, whereas a similar beaker of diethyl phthalate will not. The difference in behavior is due to the fact that the ether has a much lower flash point. The flash point is the lowest temperature, as determined by standard tests, at which a liquid gives off vapor in sufficient concentration to form an ignitable mixture with air near the surface of the liquid within the test vessel. Many common laboratory solvents and chemicals have flash points that are lower than room temperature.

Ignition Temperature

The ignition temperature (autoignition temperature) of a substance, whether solid, liquid, or gaseous, is the minimum temperature required to initiate or cause self-sustained combustion independent of the heat source. A steam line or a glowing light bulb may ignite carbon disulfide (ignition temperature 80°C). Diethyl ether (ignition temperature 160°C) can be ignited by the surface of a hot plate.

Limits of Flammability

It is possible for a flammable liquid to be above its flash point and yet not ignite in the presence of an adequate energy source. The explanation for this phenomenon lies in the composition of a fuel-air mixture that may be too lean or too rich for combustion.

Each flammable gas and liquid (as a vapor) has two fairly definite limits defining the range of concentrations in mixtures with air that will propagate flame and explode.

The *lower flammable limit* [lower explosive limit (LEL)] is the minimum concentration (percent by volume) of the vapor in air below which a flame is not propagated when an ignition source is present. Below this concentration, the mixture is too lean to burn. The *upper flammable limit* [upper explosive limit (UEL)] is the maximum concentration (percent by volume) of the vapor in air above which a flame is not propagated. Above this concentration, the mixture is too rich to burn. The flammable range (explosive range) consists of all concentrations between the LEL and the UEL. This range becomes wider with increasing temperature and in oxygen-rich atmospheres.

Because of its extreme flammability, ether is available for laboratory use only in metal containers. Carbon disulfide is almost as hazardous. The limitations of the flammability range, however, provide little margin of safety from the practical point of view because, when a solvent is spilled in the presence of an energy source, the LEL is reached very quickly and a fire or explosion will ensue before the UEL can be reached.

Spontaneous Ignition

Spontaneous ignition or combustion takes place when a substance reaches its ignition temperature without the application of external heat. The possibility of spontaneous combustion should be considered, especially when materials are stored or disposed of (see Chapters II.B, D, and E-G). Materials susceptible to spontaneous combustion include oily rags, dust accumulations, organic materials mixed with strong oxidizing agents (such as nitric acid, chlorates, permanganates, peroxides, and persulfates), alkali metals such as sodium and potassium, finely divided pyrophoric metals, and phosphorus.

The flash points, boiling points, flammable limits, and ignition temperatures of a number of common laboratory chemicals are given in Table 1 and in the Safety Data Sheets (see Section I.E.2). It should be remembered, however, that tabulations of properties of flammable substances are based on standard test methods for which the conditions may be very different from those encountered in practical use. Large safety factors should be applied. For example, the published flammable limits of vapors are for uniform mixtures with air. In a real situation, point concentrations that are much higher than the average may exist. Thus, it is good practice to set the maximum allowable concentration for safe working conditions at some fraction of the tabulated lower LEL; 20% is a commonly accepted value.

Handling Flammable Liquids

Among the most hazardous liquids are those that have flash points at room temperature or lower, particularly if their range of flammability is broad. Materials having flash points higher than the maximum ambient summer temperature do not ordinarily form ignitable mixtures with air under normal (unheated) conditions but, as shown in Table 1, many commonly used substances are potentially very hazardous, even under relatively cool conditions.

Sources of Ignition

For a fire to occur, three distinct conditions must exist simultaneously: a concentration of flammable gas or vapor that is within the flammable limits of the substance; an oxidizing atmosphere, usually air; and a source

TABLE 1 Flash Points, Boiling Points, Ignition Temperatures, and Flammable Limits of Some Common Laboratory Chemicals

Chemical	Class	Flash Point (°C)	Boiling Point (°C)	Ignition Temperature (°C)	Flammable Limit (percent by volume in air) Lower	Upper
Acetaldehyde	1A	-37.8	21.1	175.0	4.0	60.0
Acetone	1B	-17.8	56.7	465.0	2.6	12.8
Benzene	1B	-11.1	80.0	560.0	1.3	7.1
Carbon disulfide	1B	-30.0	46.1	80.0	1.3	50.0
Cyclohexane	1B	-20.0	81.7	245.0	1.3	8.0
Diethyl ether	1A	-45.0	35.0	160.0	1.9	36.0
Ethyl alcohol	1B	12.8	78.3	365.0	3.3	19.0
n-Heptane	1B	- 3.9	98.3	215.0	1.05	6.7
n-Hexane	1B	-21.7	68.9	225.0	1.1	7.5
Isopropyl alcohol	1B	11.7	82.8	398.9	2.0	12.0
Methyl alcohol	1B	11.1	64.9	385.0	6.7	36.0
Methyl ethyl ketone	1B	- 6.1	80.0	515.6	1.8	10.0
Pentane	1A	-40.0	36.1	260.0	1.5	7.8
Styrene	1B	32.2	146.1	490.0	1.1	6.1
Toluene	1B	4.4	110.6	480.0	1.2	7.1
p-Xylene	1C	27.2	138.3	530.0	1.1	7.0

of ignition. Removal of any of the three will prevent the start of a fire or extinguish an existing fire. In most situations, air cannot be excluded. The problem, therefore, usually resolves itself into preventing the coexistence of flammable vapors and an ignition source. Because spillage of a flammable liquid is always a possibility, strict control of ignition sources is mandatory.

Many sources—electrical equipment (see Chapter III.D), open flames, static electricity, burning tobacco, lighted matches, and hot surfaces—can cause ignition of flammable substances. When these materials are used in the laboratory, close attention should be given to all potential sources of ignition in the vicinity. The vapors of all flammable liquids are heavier than air and capable of traveling considerable distances. This possibility should be recognized, and special note should be taken of ignition sources at a lower level than that at which the substance is being used.

Flammable vapors from massive sources such as spillages have been known to descend into stairwells and elevator shafts and ignite on a lower story. If the path of vapor within the flammable range is continuous, the flame will propagate itself from the point of ignition back to its source.

Metal lines and vessels discharging flammable substances should be properly bonded and grounded to discharge static electricity. When nonmetallic containers (especially plastic) are used, the bonding can be made to the liquid rather than to the container. When no solution to the static problem can be found, then all processes should be carried out as slowly as possible to give the accumulated charge time to disperse.

Use of Flammable Substances

The basic precautions for safe handling of flammable materials include the following:

1. Flammable substances should be handled only in areas free of ignition sources.
2. Flammable substances should never be heated by using an open flame. Preferred heat sources include steam baths, water baths, oil baths, heating mantles, and hot air baths (see Section I.G.6).
3. When transferring flammable liquids in metal equipment, static-generated sparks should be avoided by bonding and the use of ground straps (see Section II.D.2).
4. Ventilation is one of the most effective ways to prevent the formation of flammable mixtures. An exhaust hood should be used whenever appreciable quantities of flammable substances are transferred from one container to another, allowed to stand in open containers, heated in open containers, or handled in any other way.

Flammable or Explosive Gases and Liquefied Gases

Compressed or liquefied gases present hazards in the event of fire because the heat will cause the pressure to increase and may rupture the container. Leakage or escape of flammable gases can produce an explosive atmosphere in the laboratory. Acetylene, hydrogen, ammonia, hydrogen sulfide, and carbon monoxide are especially hazardous. Acetylene and hydrogen have very wide flammability limits, which adds greatly to their potential fire and explosion hazard. (See Section I.D.1 for specific precautions for the use of flammable gases.)

Even if it is not under pressure, a substance is more concentrated in the form of a liquified gas than in the vapor phase and may evaporate extremely rapidly. Oxygen, in particular, is an extreme hazard; liquefied air is almost as dangerous [if allowed to boil freely, it will have an increasing concentration of oxygen (bp −183°C) because the nitrogen (bp −196°C) will boil away first]; and even liquid nitrogen, if it has been standing around for some time, will have absorbed enough oxygen to require careful handling. When a liquefied gas is used in a closed system, pressure may build up, so that adequate venting is required. If the liquid is flammable (e.g., hydrogen), explosive concentrations may develop. Any, or all, of the three problems, flammability, toxicity, and pressure buildup, may become serious (see Section I.D.2).

Dusts

Suspensions of oxidizable particles (such as magnesium powder, zinc dust, or flowers of sulfur) in the air constitute a powerful explosive mixture. Care should be exercised in handling these materials to avoid exposure to ignition sources.

I.C.2 HIGHLY REACTIVE CHEMICALS AND EXPLOSIVES

When the term "routine chemical reaction" is used, it usually implies a safe reaction—safe because the reaction rate is relatively slow or can be easily controlled. Highly reactive chemicals can lead to reactions that differ from routine mainly in the rate at which they progress. Reaction rates almost always increase rapidly as the temperature increases. If the heat evolved in a reaction is not dissipated, the reaction rate can increase until an explosion results. This factor must be considered, particularly when scaling up experiments, so that sufficient cooling and surface for heat exchange can be provided.

Some chemicals decompose when heated. Slow decomposition may not be noticeable on a small scale but on a large scale, or if the evolved heat and gases are confined, an explosive situation can develop. The heat-initiated decomposition of some substances, such as certain peroxides, is almost instantaneous.

Light, mechanical shock, and certain catalysts are also initiators of explosive reactions. Hydrogen and chlorine react explosively in the presence of light. Examples of shock-sensitive materials include acetylides, azides, organic nitrates, nitro compounds, and many peroxides. Acids, bases, and other substances catalyze the explosive polymerization of acrolein, and many metal ions can catalyze the violent decomposition of hydrogen peroxide.

Not all explosions result from chemical reactions. A dangerous, physically caused explosion can occur if a hot liquid (such as oil) is brought into sudden contact with a lower-boiling-point one (such as water). The instantaneous vaporization of the lower-boiling-point substance can be hazardous to personnel and destructive to equipment.

Any given sample may be just atypical enough (by virtue of impurities and such) to be dangerous. Furthermore, the hazard is associated not with the total energy released, but rather with the amazingly high rate of a detonation reaction. A high-order explosion of even milligram quantities can drive small fragments of glass or other matter deep into the body.

Organic Peroxides

Organic peroxides are a special class of compounds that have unusual stability problems that make them among the most hazardous substances normally handled in laboratories. As a class, they are low-power explosives, hazardous because of their extreme sensitivity to shock, sparks, or other forms of accidental ignition. Many peroxides that are routinely handled in laboratories are far more sensitive to shock than most primary explosives (e.g., TNT). Peroxides have a specific half-life, or rate of decomposition, under any given set of conditions. A low rate of decomposition may autoaccelerate and cause a violent explosion, especially in bulk quantities of peroxide. These compounds are sensitive to heat, friction, impact, and light, as well as to strong oxidizing and reducing agents. All organic peroxides are highly flammable, and fires involving bulk quantities of peroxides should be approached with extreme caution. A peroxide present as a contaminant in a reagent or solvent can change the course of a planned reaction.

Types of compounds known to form peroxides include the following:

1. Aldehydes.
2. Ethers, especially cyclic ethers and those containing primary and secondary alcohol groups, form dangerously explosive peroxides on exposure to air and light. [Several acceptable colorimetric tests for peroxides in ether (e.g., Jorissen reagent) are available. If a test is positive, the contaminated liquid can be filtered through a column of chromatographic, basic grade, aluminum oxide until the test is negative. The contaminated alumina should be discarded promptly. Ethers must never be distilled unless known to be free of peroxides. Containers of diethyl or diisopropyl ether should be dated when they are opened. If they are still in the laboratory after 1 month, the ether should be tested for peroxides before use. If peroxides are found, the material should be decontaminated or destroyed.]
3. Compounds containing benzylic hydrogen atoms—Such compounds are especially susceptible to peroxide formation if the hydrogens are on tertiary carbon atoms [e.g., cumene (isopropyl benzene)].
4. Compounds containing the allylic ($CH_2{=}CHCH_2R$) structure, including most alkenes.

5. Vinyl and vinylidene compounds, (e.g., vinyl acetate and vinylidene chloride).

Specific chemicals that can form dangerous concentrations of peroxides on long exposure to air are cyclohexene, cyclooctene, decalin (decahydronaphthalene), *p*-dioxane, diethyl ether, diisopropyl ether, tetrahydrofuran (THF), and tetralin (tetrahydronaphthalene).

Precautions for handling peroxides include the following:

1. The quantity of peroxide should be limited to the minimum amount required. Unused peroxides should not be returned to the container.
2. All spills should be cleaned up immediately. Solutions of peroxides can be absorbed on vermiculite.
3. The sensitivity of most peroxides to shock and heat can be reduced by dilution with inert solvents, such as aliphatic hydrocarbons. However, toluene is known to induce the decomposition of diacyl peroxides.
4. Solutions of peroxides in volatile solvents should not be used under conditions in which the solvent might be vaporized because this will increase the peroxide concentration in the solution.
5. Metal spatulas should not be used to handle peroxides because contamination by metals can lead to explosive decomposition. Ceramic or wooden spatulas may be used.
6. Smoking, open flames, and other sources of heat should not be permitted near peroxides.
7. Friction, grinding, and all forms of impact should be avoided near peroxides (especially solid ones). Glass containers that have screw-cap lids or glass stoppers should not be used. Polyethylene bottles that have screw-cap lids may be used.
8. To minimize the rate of decomposition, peroxides should be stored at the lowest possible temperature consistent with their solubility or freezing point. Liquid or solutions of peroxides should not be stored at or lower than the temperature at which the peroxide freezes or precipitates because peroxides in these forms are extremely sensitive to shock and heat.

Disposal of Peroxides

Pure peroxides should never be disposed of directly. Peroxides must be diluted before disposal.

Small quantities (25 g or less) of peroxides are generally disposed of by dilution with water to a concentration of 2% or less and then transfer of the solution to a polyethylene bottle containing an aqueous solution of a reducing agent, such as ferrous sulfate or sodium bisulfite. The material can then be handled like any other waste chemical; however, it must not be mixed with other chemicals for disposal (see Chapter II.E). Spilled peroxides should be absorbed on vermiculite as quickly as possible. The vermiculite-peroxide mixture can be burned directly or may be stirred with a suitable solvent to form a slurry that can be treated as described above. Organic peroxides should never be flushed down the drain.

Large quantities (more than 25 g) of peroxides require special handling. Each case shoula be considered separately, and handling, storage, and disposal procedures should be determined by the physical and chemical properties of the particular peroxide (see Chapters II.E and G). [See also Hamstead, A.C. Ind. Eng. Chem. 1964 56(6), 37 and Safety Data Sheet on Diethyl Ether (Section I.E).]

Some Additional Explosive Compounds

A compound is apt to be explosive if its heat of formation is smaller by more than about 100 cal/g than the sum of the heats of formation of its products. In making this calculation, one should assume a reasonable reaction to yield the most exothermic products.

In general, compounds containing the following functional groups tend to be sensitive to heat and shock: acetylide, azide, diazo, halamine, nitroso, ozonide, and peroxide [e.g., diazomethane (CH_2N_2) may decompose explosively when exposed to a ground glass joint].

Compounds containing nitro groups may be highly reactive, especially if other substituents such as halogens are present. Perchlorates, chlorates, nitrates, bromates, chlorites, and iodates, whether organic or inorganic, should be treated with respect, especially at higher temperatures.

Handling Explosive Compounds

Explosive chemicals decompose under conditions of mechanical shock, elevated temperature, or chemical action with forces that release large volumes of gases, heat, toxic vapors, or combinations thereof. Various state and federal regulations exist covering the transportation, storage, and use of explosives. These regulations should be consulted before explosives and related dangerous materials are used in the laboratory.

Explosive materials should be brought into the laboratory only as required and then in the smallest quantities adequate for the experiment being conducted. Explosives should be segregated from other materials that could create a serious hazard to life or property should an accident occur (see Section II.B.4).

The handling of highly energetic substances without injury demands attention to the most minute detail. The unusual nature of work involving such substances requires special safety measures and handling techniques that must be thoroughly understood and followed by all persons involved. The practices listed below are a guide for use in any laboratory operation that might involve explosive materials.

Personal Protective Apparel (see Sections I.F.1-3)

This includes the following items:

1. Safety glasses that have a cup-type side shield made of a light, clear plastic material affixed to the frame should be worn by all laboratory personnel.
2. A face shield that has a "snap-on" throat protector in place should be worn at all times that the worker is in a hazardous, exposed position [e.g., when operating or manipulating synthesis systems, when bench shields are moved aside, or when handling or transporting such products].
3. Gloves should be worn whenever it is necessary to reach behind a shielded area while a hazardous experiment is in progress or when handling adducts or gaseous reactants [the yellow "electrical" lineman's gloves afford good protection against 2-g quantity detonations in glass provided the detonation is 7.5 cm (3 in.) away; however, such a detonation in contact with a gloved hand would cause severe injury and probable loss of fingers].
4. Laboratory coats should be worn at all times while in explosives laboratories. They should be of a slow-burning material and fitted with quick-release cloth buttons (these coats help reduce minor injuries from flying glass and reduce the possibility of injury from an explosive flash).

Protective Devices (see Section I.F.4)

Barriers such as shields, barricades, and guards should be used to protect personnel and equipment from injury or damage. The barrier should completely surround the hazardous area. On benches and hoods, a 0.25-in.-thick acrylic sliding shield effectively protects a worker from glass fragments resulting from a laboratory-scale detonation. The shield should be closed whenever hazardous reactions are in progress or whenever hazardous materials are being temporarily stored. Such shielding is not effective against metal shrapnel.

Dry boxes should be fitted with safety glass windows overlaid with 0.25-in.-thick acrylic. This protection is adequate against an internal 5-g quantity detonation. The problem of hand protection, however, still remains, although electric lineman's gloves over the rubber dry box gloves offer some additional protection (see Section I.F.2). Other safety devices, as required, should be used in conjunction with the gloves (e.g., tongs and lab-jack turners).

Armored hoods or barricades made with extra-thick (1.0 in.) polyvinyl butyral resin shielding and heavy metal walls give complete protection against detonations not in excess of the acceptable 20-g limit. These hoods are designed for use with 100 g of material, but an arbitrary 20-g limit is usually set because of the noise level in the event of a detonation. Such hoods should be equipped with mechanical hands that enable the operator to remotely manipulate equipment and handle adduct containers. A sign, such as CAUTION: NO ONE MAY ENTER AN ARMORED HOOD FOR ANY REASON DURING THE COURSE OF A HAZARDOUS OPERATION should be posted.

Miscellaneous protective devices such as both long- and short-handled tongs for holding or manipulating hazardous items at a safe distance and remote-control equipment (e.g., mechanical arms, stopcock turners, lab-jack turners, and remote cable controllers) should be available as required to prevent exposure of any part of the body to injury.

Reaction Quantities

In conventional explosives laboratories, no more than 0.5 g of product should be prepared in a single run. During the actual reaction period, no more than 2 g of reactants should be present in the reaction vessel. This means that the diluent, the substrate, and the energetic reactant must all be considered when determining the total explosive power of the reaction mixture. Special reviews should be established to examine operational and safety problems involved in scaling up a reaction in which an explosive substance is used.

Reaction Operations

Various heating methods may be used. The most common are heating tapes and mantles and sand, water, steam, and silicone oil baths. Hair dryers (heat guns) can be used for certain operations; however, their use should be prohibited when a flammable vapor potential is present. All controls for heating and stirring equipment should be operable from outside the shielded area.

Vacuum pumps should carry tags indicating the date of the most recent oil change. Oil should be changed once a month (sooner, if it is known that the oil has been exposed to reactive gases). All pumps should either be vented into a hood or trapped. Vent lines may be Tygon®, rubber, or copper. If Tygon or rubber lines are used, they should be supported so that they do not sag (see Section I.G.2).

When potentially explosive materials are being handled, the area should be posted with a sign such as WARNING: VACATE THE AREA AT THE FIRST SIGN OF ODOR. STAY OUT UNTIL THE VENTILATION SYSTEM HAS CLEARED THE AIR.

When condensing explosive gases, the temperature of the bath and the effect on the reactant gas of the condensing material selected must be determined experimentally. Very small quantities should be used because detonations may occur. In all cases, a shielded Dewar flask (see Section I.D.3) should be used when condensing reactants. Maximum quantity limits should be observed. Heating baths of flammable materials should be prohibited.

I.C.3 OTHER SPECIFIC CHEMICAL HAZARDS THAT FREQUENTLY LEAD TO FIRES OR EXPLOSIONS

Hazardous Substances or Objects

Acetylenic compounds are explosive in mixtures of 2.5-80% with air. At pressures of 2 or more atmospheres, acetylene (C_2H_2) subjected to an electrical discharge or high temperature decomposes with explosive violence. Dry acetylides detonate on receiving the slightest shock. (See Section I.D.4.)

Aluminum chloride ($AlCl_3$) should be considered a potentially dangerous material. If moisture is present, there may be sufficient decomposition [to give hydrogen chloride (HCl)] to build up considerable pressure. If a bottle is to be opened after long standing, it should be completely enclosed in a heavy towel.

Ammonia (NH_3) reacts with iodine to give nitrogen triiodide, which is explosive, and with hypochlorites to give chlorine. Mixtures of NH_3 and organic halides sometimes react violently when heated under pressure.

Dry *benzoyl peroxide* $(C_6H_5CO_2)_2$ is easily ignited and sensitive to shock. It decomposes spontaneously at temperatures above 50°C. It is reported to be desensitized by addition of 20% water.

Carbon disulfide (CS_2) is both very toxic and very flammable; mixed with air, its vapors can be ignited by a steam bath or pipe, a hot plate, or a glowing light bulb. [See Safety Data Sheet (Section I.E.2).]

Chlorine (Cl_2) may react violently with hydrogen (H_2) or with hydrocarbons when exposed to sunlight. [See Safety Data Sheet (Section I.E.2).]

Chromium trioxide-pyridine complex ($CrO_3 \cdot C_5H_5N$) may explode if the CrO_3 concentration is too high. The complex should be prepared by addition of CrO_3 to excess C_5H_5N.

Diazomethane (CH_2N_2) and related compounds should be treated with extreme caution. They are very toxic, and the pure gases and liquids explode readily. Solutions in ether are safer from this standpoint. (See also "Organic Syntheses," Rabjohn, N., Ed.; Wiley: New York, 1963, Collective Volume IV, pp. 250-253.

Dimethyl sulfoxide [$(CH_3)_2SO$] decomposes violently on contact with a wide variety of active halogen compounds. Explosions from contact with active metal hydrides have been reported. Its toxicity is still unknown, but it does penetrate and carry dissolved substances through the skin membrane.

Dry ice should not be kept in a container that is not designed to withstand pressure. Containers of other substances stored over dry ice for extended periods generally absorb carbon dioxide (CO_2) unless they have been sealed with care. When such containers are removed from storage and allowed to come rapidly to room temperature, the CO_2 may develop sufficient pressure to burst the container with explosive violence. On removal of such containers from storage, the stopper should be loosened or the container itself should be wrapped in towels and kept behind a shield. Dry ice can produce serious burns (this is also true for all types of cooling baths).

Drying agents—Ascarite® should not be mixed with phosphorus pentoxide (P_2O_5) because the mixture may explode if it is warmed with a trace of water. Because the cobalt salts used as moisture indicators in some drying agents may be extracted by some organic solvents, the use of these drying agents should be restricted to gases.

Diethyl, diisopropyl, and *other ethers* (particularly the branched-chain type) sometimes explode during heating or refluxing because of the presence of peroxides. Ferrous salts or sodium bisulfite can be used to decompose these peroxides, and passage over basic active alumina will remove most of the peroxidic material (see Section I.C.2). In general, however, old samples of ethers should be discarded.

Ethylene oxide (C_2H_4O) has been known to explode when heated in a closed vessel. Experiments using ethylene oxide under pressure should be carried out behind suitable barricades (see Section I.C.2).

Halogenated compounds—Chloroform ($CHCl_3$), carbon tetrachloride (CCl_4), and other halogenated solvents should not be dried with sodium, potassium, or other active metal; violent explosions are usually the result of such attempts. Many halogenated compounds are toxic.

Hydrogen peroxide (H_2O_2) stronger than 3% can be dangerous; in contact with the skin, it may cause severe burns. Thirty percent H_2O_2 may decompose violently if contaminated with iron, copper, chromium, or other metals or their salts.

Liquid-nitrogen-cooled traps open to the atmosphere rapidly condense liquid air. Then, when the coolant is removed, an explosive pressure buildup occurs, usually with enough force to shatter glass equipment. Hence, only sealed or evacuated equipment should be so cooled (see Section I.D.3).

Lithium aluminum hydride ($LiAlH_4$) should not be used to dry methyl ethers or tetrahydrofuran; fires from this are very common. The products of its reaction with CO_2 have been reported to be explosive. Carbon dioxide or bicarbonate extinguishers should not be used against $LiAlH_4$ fires, which should be smothered with sand or some other inert substance.

Oxygen tanks—Serious explosions have resulted from contact between oil and high-pressure oxygen. Oil should not be used on connections to an O_2 cylinder (see Section I.D.4).

Ozone (O_3) is a highly reactive and toxic gas. It is formed by the action of ultraviolet light on oxygen (air) and, therefore, certain ultraviolet sources may require venting to the exhaust hood.

Palladium or *platinum on carbon, platinum oxide, Raney nickel,* and *other catalysts* should be filtered from catalytic hydrogenation reaction mixtures carefully. The recovered catalyst is usually saturated with hydrogen and highly reactive and, thus, will enflame spontaneously on exposure to air. Particularly in large-scale reactions, the filter cake should not be allowed to become dry. The funnel containing the still-moist catalyst filter cake should be put into a water bath immediately after completion of the filtration.

Another hazard in working with such catalysts is the danger of explosion if additional catalyst is added to a flask in which hydrogen is present.

Parr bombs used for hydrogenations have been known to explode. They should be handled with care behind shields, and the operator should wear goggles.

Perchlorates—The use of perchlorates should be avoided wherever possible. Perchlorates should not be used as drying agents if there is a possibility of contact with organic compounds or in proximity to a dehydrating acid strong enough to concentrate the perchloric acid ($HClO_4$) to more than 70% strength (e.g., in a drying train that has a bubble counter containing sulfuric acid). Safer drying agents should be used.

Seventy percent $HClO_4$ can be boiled safely at approximately 200°C, but contact of the boiling undiluted acid or the hot vapor with organic matter, or even easily oxidized inorganic matter (such as compounds of trivalent antimony), will lead to serious explosions. Oxidizable substances must never be allowed to contact $HClO_4$. Beaker tongs, rather than rubber gloves, should be used when handling fuming $HClO_4$. Perchloric acid evaporations should be carried out in a hood that has a good draft. Frequent (weekly) washing out of the hood and ventilator ducts with water is necessary to avoid danger of spontaneous combustion or explosion if this acid is in common use.

Permanganates are explosive when treated with sulfuric acid. When both compounds are used in an absorption train, an empty trap should be placed between them.

Peroxides (inorganic)—When mixed with combustible materials, barium, sodium, and potassium peroxides form explosives that ignite easily.

Phosphorus (P) (red and white) forms explosive mixtures with oxidizing agents. White P should be stored under water because it is spontaneously flammable in air. The reaction of P with aqueous hydroxides gives phosphine, which may ignite spontaneously in air or explode.

Phosphorus trichloride (PCl_3) reacts with water to form phosphorous acid, which decomposes on heating to form phosphine, which may ignite spontaneously or explode. Care should be taken in opening containers of PCl_3, and samples that have been exposed to moisture should not be heated without adequate shielding to protect the operator.

Potassium (K) is in general more reactive than sodium; it ignites quickly on exposure to humid air and, therefore, should be handled under the surface of a hydrocarbon solvent such as mineral oil or toluene (see *Sodium*).

Residues from vacuum distillations (for example, ethyl palmitate) have been known to explode when the still was vented to the air before the residue was cool. Such explosions can be avoided by venting the still pot with nitrogen, by cooling it before venting, or by restoring the pressure slowly.

Sodium (Na) should be stored in a closed container under kerosene, toluene, or mineral oil. Scraps of Na or K should be destroyed by reaction with *n*-butyl alcohol. Contact with water should be avoided because Na reacts violently with water to form H_2 with evolution of sufficient heat to cause ignition. Neither carbon dioxide or bicarbonate nor carbon tetrachloride fire extinguishers should be used on alkali metal fires.

Sulfuric acid (H_2SO_4) should be avoided, if possible, as a drying agent in desiccators. If it must be used, glass beads should be placed in it to help prevent splashing when the desiccator is moved. The use of H_2SO_4 in melting point baths should be avoided. (Silicone oil should be used.) To dilute H_2SO_4, add the acid slowly to cold water.

Trichloroethylene (Cl_2CCHCl) reacts under a variety of conditions with potassium or sodium hydroxide to form dichloroacetylene, which ignites spontaneously in air and detonates readily even at dry-ice temperatures. The compound itself is highly toxic, and suitable precautions should be taken when it is used as a degreasing solvent. [See Safety Data Sheet (Section I.E.2).]

Incompatible Chemicals

When transporting, storing (see Chapters II.A and B), using, or disposing of any substance, utmost care must be exercised to ensure that the substance cannot accidently come in contact with another with which it is incompatible. Such contact could result in a serious explosion or the formation of substances that are highly toxic or flammable or both.

Table 2 is a guide to avoiding accidents involving incompatible substances.

TABLE 2 Examples of Incompatible Chemicals

Chemical	Is Incompatible With
Acetic acid	Chromic acid, nitric acid, hydroxyl compounds, ethylene glycol, perchloric acid, peroxides, permanganates
Acetylene	Chlorine, bromine, copper, fluorine, silver, mercury
Acetone	Concentrated nitric and sulfuric acid mixtures
Alkali and alkaline earth metals (such as powdered aluminum or magnesium, calcium, lithium, sodium, potassium)	Water, carbon tetrachloride or other chlorinated hydrocarbons, carbon dioxide, halogens
Ammonia (anhydrous)	Mercury (in manometers, for example), chlorine, calcium hypochlorite, iodine, bromine, hydrofluoric acid (anhydrous)
Ammonium nitrate	Acids, powdered metals, flammable liquids, chlorates, nitrites, sulfur, finely divided organic or combustible materials
Aniline	Nitric acid, hydrogen peroxide
Arsenical materials	Any reducing agent
Azides	Acids
Bromine	See Chlorine
Calcium oxide	Water
Carbon (activated)	Calcium hypochlorite, all oxidizing agents
Carbon tetrachloride	Sodium
Chlorates	Ammonium salts, acids, powdered metals, sulfur, finely divided organic or combustible materials
Chromic acid and chromium trioxide	Acetic acid, naphthalene, camphor, glycerol, alcohol, flammable liquids in general
Chlorine	Ammonia, acetylene, butadiene, butane, methane, propane (or other petroleum gases), hydrogen, sodium carbide, benzene, finely divided metals, turpentine
Chlorine dioxide	Ammonia, methane, phosphine, hydrogen sulfide
Copper	Acetylene, hydrogen peroxide
Cumene hydroperoxide	Acids (organic or inorganic
Cyanides	Acids
Flammable liquids	Ammonium nitrate, chromic acid, hydrogen peroxide, nitric acid, sodium peroxide, halogens
Fluorine	Everything
Hydrocarbons (such as butane, propane, benzene)	Fluorine, chlorine, bromine, chromic acid, sodium peroxide
Hydrocyanic acid	Nitric acid, alkali
Hydrofluoric acid (anhydrous)	Ammonia (aqueous or anhydrous)
Hydrogen peroxide	Copper, chromium, iron, most metals or their salts, alcohols, acetone, organic materials, aniline, nitromethane, combustible materials
Hydrogen sulfide	Fuming nitric acid, oxidizing gases
Hypochlorites	Acids, activated carbon
Iodine	Acetylene, ammonia (aqueous or anhydrous), hydrogen
Mercury	Acetylene, fulminic acid, ammonia

TABLE 2 (*continued*)

Chemical	Is Incompatible With
Nitrates	Sulfuric acid
Nitric acid (concentrated)	Acetic acid, aniline, chromic acid, hydrocyanic acid, hydrogen sulfide, flammable liquids, flammable gases, copper, brass, any heavy metals
Nitrites	Acids
Nitroparaffins	Inorganic bases, amines
Oxalic acid	Silver, mercury
Oxygen	Oils, grease, hydrogen, flammable liquids, solids, or gases
Perchloric acid	Acetic anhydride, bismuth and its alloys, alcohol, paper, wood, grease, oils
Peroxides, organic	Acids (organic or mineral), avoid friction, store cold
Phosphorus (white)	Air, oxygen, alkalis, reducing agents
Potassium	Carbon tetrachloride, carbon dioxide, water
Potassium chlorate	Sulfuric and other acids
Potassium perchlorate (see also chlorates)	Sulfuric and other acids
Potassium permanganate	Glycerol, ethylene glycol, benzaldehyde, surfuric acid
Selenides	Reducing agents
Silver	Acetylene, oxalic acid, tartartic acid, ammonium compounds, fulminic acid
Sodium	Carbon tetrachloride, carbon dioxide, water
Sodium nitrite	Ammonium nitrate and other ammonium salts
Sodium peroxide	Ethyl or methyl alcohol, glacial acetic acid, acetic anhydride, benzaldehyde, carbon disulfide, glycerin, ethylene glycol, ethyl acetate, methyl acetate, furfural
Sulfides	Acids
Sulfuric acid	Potassium chlorate, potassium perchlorate, potassium permanganate (similar compounds of light metals, such as sodium, lithium)
Tellurides	Reducing agents

I.D

Procedures for Working with Compressed Gases and for Working at Pressures Above or Below Atmospheric

Compressed gases present a unique hazard in that they have the potential for simultaneous exposure to both mechanical and chemical hazards (depending on the particular gas). If the gas is flammable, flash points lower than room temperature compounded by high rates of diffusion (instant permeation throughout the laboratory) present the danger of fire or explosion. Additional hazards can arise from the reactivity and toxicity of the gas, and asphyxiation can be caused by high concentrations of even "harmless" gases such as nitrogen. Finally, the large amount of potential energy resulting from compression of the gas [some gases are compressed to pressures up to 41 MPa (6000 lbf/in.^{2})] makes a compressed gas cylinder a potential rocket or fragmentation bomb.

Thus, careful procedures are necessary for handling the various types of compressed gases, the cylinders that contain them, the regulators used to control their flow, and the piping used to confine them during flow.

I.D.1 CYLINDERS OF COMPRESSED GASES

A compressed gas as defined by the U.S. Department of Transportation (DOT), which regulates their commercial transport, is "any material or mixture having in the container either an absolute pressure greater than 276 kPa (40 lbf/in.^{2}) at 21°C, or an absolute pressure greater than 717 kPa (104 lbf/in.^{2}) at 54°C, or both, or any liquid flammable material having a Reid vapor pressure greater than 276 kPa (40 lbf/in.^{2}) at 38°C." The DOT also has established codes that specify the materials to be used for the construction and the capacities, test procedures, and service pressures of the cylinders in which compressed gases are stored. A typical compressed gas steel cylinder might be designated DOT 3A-2000, which indicates that it has been manufactured under specification 3A and has an operating pressure of 13.8 MPa (2000 lbf/in.^{2}) at 21°C. However, regardless of the pressure rating of the cylinder, the physical state of the material within it determines the pressure of the gas. For example, liquefied gases such as propane or ammonia will exert their own vapor pressure as long as any liquid remains in the cylinder and the critical temperature is not exceeded.

Prudent procedures for the use of compressed gas cylinders in the laboratory include attention to proper identification, transportation and storage (see Chapters II.B and C), handling and use, and return of the empty cylinder.

Identification

The contents of any compressed gas cylinder should be clearly identified so as to be easily, quickly, and completely determined by any laboratory worker. Such identification should be stenciled or stamped on the cylinder itself or a label should be provided that cannot be removed from the cylinder (three-part tag systems, which are available commercially, can be very useful for identification and inventory). No compressed gas cylinder should be accepted for use that does not legibly identify its contents by name (see Chapter II.A). Color coding is not a reliable means of identification; cylinder colors vary from supplier to supplier and labels on caps have no value as caps are interchangeable. If the labeling on a cylinder becomes unclear or an attached tag is defaced, so that the contents cannot be identified, the cylinder should be marked "contents unknown" and returned directly to the manufacturer.

All gas lines leading from a compressed gas supply should be clearly labeled so as to identify the gas, the laboratory served, and relevant emergency telephone numbers. The labels should be color coded to distinguish hazardous gases (such as flammable, toxic, or corrosive substances) (e.g., a yellow background and black letters) from safe (inert) gases (e.g., a green background and black letters). Signs should be conspicuously posted in areas in which flammable compressed gases are stored, identifying the substances and appropriate precautions (e.g., HYDROGEN-FLAMMABLE GAS—NO SMOKING—NO OPEN FLAMES).

Handling and Use of Cylinders

Compressed gas cylinders should be firmly secured at all times. A clamp and belt or chain are generally suitable for this purpose. Pressure-relief devices protecting equipment that is attached to cylinders of flammable, toxic, or otherwise hazardous gases should be vented to a safe place.

Standard cylinder-valve outlet connections have been devised by the Compressed Gas Association (CGA) to prevent the mixing of incompatible gases due to an interchange of connections. The outlet threads used vary in diameter; some are internal and some are external; some are right-handed and some are left-handed. In general, right-handed threads are used for nonfuel and water-pumped gases, and left-handed threads are used for fuel and oil-pumped gases. Information on the standard equipment assemblies for use with specific compressed gases are available from the supplier. To minimize undesirable connections that may result in a hazard, only CGA standard combinations of valves and fittings should be used in compressed gas installations; the assembly of miscellaneous parts (even of standard approved types) should be avoided. The threads on cylinder valves, regulators, and other fittings should be examined to ensure that they correspond to one another and are undamaged.

Cylinders should be placed so that the cylinder valve is accessible at all times. The main cylinder valve should be closed as soon as it is no longer necessary that it be open (i.e., it should never be left open when the equipment is unattended or not operating). This is necessary not only for safety when the cylinder is under pressure, but also to prevent the corrosion and contamination that would result from diffusion of air and moisture into the cylinder after it has been emptied.

Most cylinders are equipped with hand-wheel valves. Those that are not should have a spindle key on the valve spindle or stem while the cylinder is in service. Only wrenches or other tools provided by the cylinder supplier should be used to open a valve. In no case should pliers be used to open a cylinder valve. Some valves require washers, and this should be checked before the regulator is fitted.

Cylinder valves should be opened slowly; the valve on an unregulated cylinder should never be "cracked." It is never necessary to open the main cylinder valve all the way; the resulting flow will be much greater than one would ever want. It is safe practice to open the main valve only to the extent necessary.

When opening the valve on a cylinder containing an irritating or toxic gas, the user should stand on the upwind side of the cylinder with the valve pointed downwind and should warn those working nearby in case of a possible leak.

Pressure Regulators

Pressure regulators are generally used to reduce a high-pressure supplied gas to a desirable lower pressure and to maintain a satisfactory delivery pressure and flow level for the required operating conditions. They can be obtained to fit many operating conditions in regard to range of supply and delivery pressures, flow capacity, and construction materials. All regulators are of a diaphragm type, spring-loaded or gas-loaded, depending on pressure requirements. They may be single-stage or two-stage. Under no circumstances should oil be used on regulator valves or cylinder valves.

Each regulator is supplied with a specific CGA standard inlet connection to fit the outlet connection on the cylinder valve for the particular gas (see above).

Regulators for use with noncorrosive gases are usually made of brass. Special regulators made of corrosion-resistant materials can be obtained for use with such gases as ammonia, boron trifluoride, chlorine, hydrogen chloride, hydrogen sulfide, and sulfur dioxide. Because of freeze-up and corrosion problems, regulators used with carbon dioxide gas must have special internal design features and be made of special construction materials.

All pressure regulators should be equipped with spring-loaded pressure-relief valves to protect the low-pressure side. When used on cylinders of flammable, toxic, or otherwise hazardous gases, the relief valve should be vented to a safe location. The use of internal-bleed type regulators should be avoided.

Flammable Gases

Sparks and flames should be kept from the vicinity of cylinders of flammable gases (see Section I.C.1). An open flame should never be used to detect leaks of flammable gases. Rather, soapy water should be used, except during freezing weather, when a 50% glycerine-water solution or its equivalent may be used. Connections to piping, regulators, and other appliances should always be kept tight to prevent leakage, and the hoses used should be kept in good condition.

Regulators, hoses, and other appliances used with cylinders of flammable gases should not be interchanged with similar equipment intended for use with other gases.

All cylinders containing flammable gases should be stored in a well-ventilated place. Reserve stocks of such cylinders should never be stored in the vicinity of cylinders containing oxygen (see Section II.B.6).

EMPTY CYLINDERS

A cylinder should never be emptied to a pressure lower than 172 kPa (25 lbf/in.^{2}) because the residual contents may become contaminated if the valve is left open. Empty cylinders should not be refilled. Rather, the regulator should be removed and the valve cap should be replaced. The cylinder should be clearly marked as "empty" and returned to a storage area for pickup by the supplier. Empty and full cylinders should not be stored in the same place.

Cylinder discharge lines should be equipped with approved check valves to prevent inadvertent contamination of cylinders that are connected to a closed system where the possibility of flow reversal exists. Sucking back is particularly troublesome in the case of gases used as reactants in a closed system. A cylinder in such a system should be shut off and removed from the system when the pressure remaining in the cylinder is at least 172 kPa. If there is a possibility that a cylinder has been contaminated, it should be so labeled and returned to the supplier.

I.D.2 NON-VENDOR-OWNED PRESSURE VESSELS AND OTHER EQUIPMENT

RECORDS, INSPECTION, AND TESTING

Each pressure vessel should have stamped on it (or on an attached plate), its basic allowable working pressure, the allowable temperature at this pressure, and the material of construction. Similarly, the relieving pressure and setting data should be stamped on a metal tag attached to installed pressure-relieving devices, and the setting mechanisms should be sealed. Relief devices used on pressure regulators do not require these seals or numbers.

All pressure equipment should be tested or inspected periodically. The interval between tests or inspections is determined by the severity of the service involved. Corrosive or otherwise hazardous service will require more frequent tests and inspections. Inspection data should be stamped on or attached to the equipment.

The use of soap solution and air or nitrogen pressure to the maximum allowable working pressure of the weakest section of the assembled apparatus is usually an adequate means of testing a system for leaks through threaded joints, packings, valves, and such.

Before any pressure equipment is altered, repaired, stored, or shipped it should be vented and all toxic or other hazardous material removed completely so it can be handled safely. Especially hazardous materials may require special cleaning techniques.

Assembly and Operation

During the assembly of pressure equipment and piping, only appropriate components should be used and care should be taken to avoid strains and concealed fractures resulting from the use of improper tools or excessive force. Piping should not be used to support equipment of any significant weight.

Threads that do not fit accurately should not be forced. Thread connections should match correctly; tapered pipe threads cannot be joined with parallel machine threads. A suitable thread lubricant should be used when assembling the apparatus. However, oil or lubricant must never be used on any equipment that will be used with oxygen. Parts having damaged or partly stripped threads should be rejected.

In assembling copper-tubing installations, sharp bends should be avoided and considerable flexibility should be allowed. Copper-tubing work hardens and cracks on repeated bending. It should be inspected frequently and renewed when necessary.

Stuffing boxes and gland joints are a likely source of trouble in pressure installations. Particular attention should be given to the proper installation and maintenance of these parts, including the proper choice of lubricant and packing material.

Experiments carried out in closed systems and involving highly reactive materials, such as those subject to rapid polymerization (e.g., dienes or unsaturated aldehydes, ketones, or alcohols) should be preceded by small-scale tests using the exact reaction materials to determine the possibility of an unexpectedly rapid reaction or unforeseen side reactions. All reactions under pressure should be shielded (see Section I.F.4).

Autoclaves and other pressure-reaction vessels should not be filled more than half full to ensure that space remains for expansion of the liquid when it is heated. Leak corrections or adjustments to the apparatus should not be made while it is pressurized; rather, the system should be depressurized before mechanical adjustments are made.

Immediately after an experiment in which low-pressure equipment connected to a source of high pressure is concluded, the low-pressure equipment should either be disconnected entirely or left independently vented to the atmosphere. This will prevent the gradual buildup of excessive pressure in the low-pressure equipment due to leakage from the high-pressure side.

Vessels or equipment made partly or entirely of silver, copper, or alloys containing more than 50% copper should not be used in contact with acetylene or ammonia. Those made of metals susceptible to amalgamation (such as copper, brass, zinc, tin, silver, lead, and gold) should not come into contact with mercury. This includes equipment that has soldered and brazed joints.

Prominent warning signs should be placed in any area where a pressure reaction is in progress so that others entering the area will be aware of the potential hazard.

Pressure-Relief Devices

All pressure or vacuum systems and all vessels that may be subjected to pressure or vacuum should be protected by pressure-relief devices. Experiments involving highly reactive materials that might explode may also require the use of special pressure-relief devices, operating at a fraction of the permissible working pressure of the system.

Examples of pressure-relief devices include the rupture-disk type used with closed-system vessels and the spring-loaded relief valves used with vessels for transferring liquefied gases. The following considerations are advisable in the use of pressure-relief devices:

1. The maximum setting of a pressure-relief device is the rated working pressure established for the vessel or for the weakest member of the pressure system at the operating temperature. The operating pressure should be less than the allowable working pressure of the system. In the case of a system protected by a spring-loaded relief device, the maximum operating pressure should be from 5 to 25% lower than the rated working pressure, depending on the type of safety valve and the importance of leak-free operation. In the case of a system protected by a rupture-disk device, the maximum operating pressure should be about two-thirds of the rated working pressure; the exact figure is governed by the fatigue life of the disk used, the temperature, and load pulsations.
2. Pressure-relief devices that may discharge toxic, corrosive, flammable, or otherwise hazardous or noxious materials should be vented in a safe and environmentally acceptable manner (see Section I.H.3).
3. Shutoff valves should not be installed between pressure-relief devices and the equipment they are to protect.
4. Only qualified persons should perform maintenance work on pressure-relief devices.
5. Pressure-relief devices should be inspected periodically.

PRESSURE GAGES

The proper choice and use of a pressure gage is the responsibility of the user. Among the factors to be considered are the flammability, compressibility, corrosivity, toxicity, temperature, and pressure range of the fluid with which it is to be used.

A pressure gage is normally a weak point in any pressure system because its measuring element must operate in the elastic zone of the metal involved. The resulting limited factor of safety makes careful gage selection and use mandatory and often dictates the use of accessory protective equipment. The primary element of the most commonly used gages is a Bourdon tube, which is usually made of brass or bronze and has soft-soldered connections. More expensive gages can be obtained that have Bourdon tubes made of steel, stainless steel, or other special metals and welded or silver-soldered connections. Accuracies vary from $\pm 2\%$ for less expensive pressure gages to $\pm 0.1\%$ for higher quality gages (these tolerances apply to the middle half or three quarters of the gage range).

Consideration should be given to alternative methods of pressure measurement that may provide greater safety than the direct use of pressure gages. Such methods include the use of seals or other isolating devices in pressure tap lines, indirect-observation devices, and remote measurement by strain-gage transducers.

GLASS EQUIPMENT

The use of glassware for work at pressure extremes should be avoided whenever possible. Glass is a brittle material subject to unexpected failures due to factors such as mechanical impact and assembly and tightening stresses. Glass equipment, such as rotameters and liquid-level gages, that is incorporated in metallic pressure systems should be installed with shutoff valves at both ends to control the discharge of liquid or gaseous materials in the event of breakage.

Glass equipment in pressure or vacuum service should be provided with adequate shielding to protect users and others in the area from flying glass and the contents of the equipment. New or repaired glass equipment for pressure or vacuum work should be examined for flaws and strains under polarized light.

Corks, rubber stoppers, and rubber or plastic tubing should not be relied on as relief devices for protection of glassware against excess pressure; a liquid seal, Bunsen tube, or equivalent positive relief device should be used. When glass pipe is used, only proper metal fittings should be used.

Plastic Equipment

Except as noted below, the use of plastic equipment for pressure or vacuum work should be avoided unless no suitable substitute is available.

Tygon and similar plastic tubing have some limited applications in pressure work. These materials can be used for natural gas, hydrocarbons, and most aqueous solutions at room temperature and moderate pressure. Details of permissible operating conditions must be obtained from the manufacturer. Because of their very large coefficients of thermal expansion, some polymers have a tendency to expand a great deal on heating. Thus, if the valve or joint is tightened when the apparatus is cold, the plastic can entirely close an opening when the temperature increases. This problem can be a hazard in equipment subjected to very low temperatures or to alternating low and high temperatures.

Piping, Tubing, and Fittings

All-brass and stainless steel fittings should be used with copper or brass and steel or stainless steel tubings, respectively. It is very important that fittings of this type be installed correctly. It is not usually advisable to mix different types of fittings in the same apparatus assembly.

Sealed-Tube Reactions

For any reaction run on a large scale (more than 10-20 g total weight of reactants) or at a maximum pressure in excess of 690 kPa (100 $lbf/in.^2$), only procedures involving a suitable high-pressure autoclave should be used. However, it is sometimes convenient to run small-scale reactions at low pressures in a small sealed glass tube or in a thick-walled pressure bottle of the type used for catalytic hydrogenation. For any such reaction, the laboratory worker should be fully prepared for the not uncommon possibility that the sealed vessel will burst. Every precaution should be taken to avoid injury from flying glass or from corrosive or toxic reactants. Centrifuge bottles should be sealed with rubber stoppers clamped in place, wrapped with friction tape, surrounded by multiple layers of loose cloth toweling, and clamped behind a good safety shield. The preferred source of heat for such vessels is steam, because an explosion in the vicinity of an electrical heater could start a fire and an explosion in a liquid heating bath would distribute hot liquid around the area. Any reaction of this type should be labeled with signs that indicate the contents of the reaction vessel and the explosion hazard.

For reactions run in sealed tubes, similar precautions should be

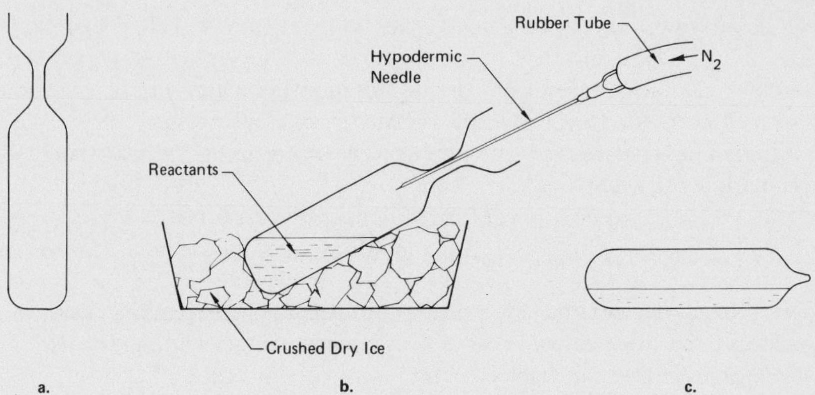

FIGURE 3 Procedure for conducting sealed-tube reactions.

followed. The sealed glass tubes can be placed inside pieces of brass or iron pipe capped at one end with a pipe cap or, alternatively, in an autoclave containing some of the reaction solvent (to equalize the pressure inside and outside the glass tube). The tubes can be heated with steam or in a specially constructed, electrically heated "sealed-tube" furnace that is thermostatically controlled and located such that if an explosion should occur its force is directed into a safe area.

Small tubes can be made of ordinary Pyrex® glass tubing having an outside diameter of 5-8 mm. Special thick-walled Pyrex pressure tubing must be used for larger-diameter reaction vessels. The reaction tube can be constructed as shown in Figure 3a if care is taken to avoid thin spots in the bottom or in the constricted section. The reactants should be added to this tube, which should never be more than one-half full (heated liquids are essentially incompressible), by using a pipet or syringe in such a way that no material is deposited on the constricted section of the tube. The tube is then swept with nitrogen (see Figure 3b) and concurrently cooled in crushed dry ice (NOT a dry ice-acetone bath). When the reactants are thoroughly chilled, the nitrogen line is removed and the constricted section is immediately sealed by using a small oxygen-gas torch. The sealed tube (Figure 3c) should be heated by steam or in a "sealed-tube" furnace.

When the required heating has been completed, the sealed tube or bottle should be allowed to cool to room temperature. If a sealed-tube furnace has been used and the contents of the tube are nonflammable, the tube may be opened by heating the tip protruding from the furnace with a small oxygen-gas flame until the pressure is released by breakage of the molten glass. Sealed bottles and tubes of flammable materials should be wrapped with cloth toweling, placed behind a safety shield, and then slowly cooled, first in an ice bath and then in dry ice, after which the clamps and rubber stoppers can be removed from the bottles and the tips of tubes heated to the melting point to release any remaining internal pressure. After the pressure has been released, the tubes can be cut open by cracking along a file mark in the usual way.

Handling of Liquefied Gases and Cryogenic Liquids

The primary hazards of cryogenic liquids are fire or explosion, pressure buildup, embrittlement of structural materials, contact with and destruction of living tissue, and asphyxiation.

The fire or explosion hazard is obvious when gases such as hydrogen, methane, and acetylene are used. Enriched oxygen will greatly increase the flammability of ordinary combustible materials and may even cause some noncombustible materials (such as carbon steel) to burn readily (see Section III.G.1).

Oxygen-saturated wood and asphalt have been known to literally explode when subjected to shock. Because it has a higher boiling point (−183°C) than nitrogen (−195°C), helium (−269°C), or hydrogen (−252.7°C), oxygen can be condensed out of the atmosphere during the use of these lower-boiling cryogenic liquids. Particularly with liquid hydrogen, conditions may exist for an explosion.

Even very brief skin contact with a cryogenic liquid is capable of causing tissue damage similar to that of thermal burns, and prolonged contact may result in blood clots that have potentially very serious consequences.

Eye protection, preferably a face shield, should be worn when handling liquefied gases and other cryogenic fluids (see Section I.F.1). Gloves should be chosen that are impervious to the fluid being handled and loose enough to be tossed off easily (see Section I.F.2). A potholder may be a desirable alternative. The area should be well ventilated. The transfer of liquefied gases from one container to another should not be attempted for the first time without the direct supervision and instruction of someone experienced in this operation.

It is advisable to furnish all cylinders and equipment used to contain liquefied gases that are not vendor-owned with a spring-loaded pressure-relief device (not a rupture-disk) because of the magnitude of the potential hazard represented by the large amount of flammable or toxic (or both) gas that can result from activation of a non-resetting relief device on a container of liquefied gas. Commercial cylinders of liquefied gases are normally supplied equipped only with a fusible-plug type of relief device, as permitted by DOT regulations.

Cylinders and other pressure vessels used for the storage and handling of liquefied gases should not be filled to more than 80% of capacity. This is a precaution against possible thermal expansion of the contents and bursting of the vessel by hydrostatic pressure. If the possibility exists that the temperature of the full cylinder might be increased to above 30°C, a lower percentage (e.g., 60) of capacity should be the limit.

Low-Temperature Equipment

At low temperatures, the impact strength of ordinary carbon steel is greatly reduced. The steel may fail when subjected to impact or mechanical shock, even though its ability to withstand slowly applied loading is not impaired. This type of failure normally occurs at points of high stress (such as at notches in the material or abrupt changes of section).

The 18% chromium-8% nickel stainless steels retain their impact resistance down to approximately −240°C, the exact value depending heavily on special design considerations. The impact resistance of aluminum, copper, nickel, and many other nonferrous metals and alloys increases with decreasing temperatures.

Hydrogen Embrittlement

Special alloy steels should be used for liquids or gases containing hydrogen at temperatures greater than 200°C or at pressures greater than 34.5 MPa (500 lbf/in.^{2}) because of the danger of weakening carbon steel equipment by hydrogen embrittlement.

I.D.3 VACUUM WORK

In an evacuated system, the higher pressure is on the outside, rather than on the inside, so that a break causes an implosion rather than an explosion. The resulting hazards consist of flying glass, spattered chemicals, and possibly fire.

A moderate vacuum, such as 10 mm Hg, which can be achieved by a water aspirator, often seems safe compared with a high vacuum, such as 10^{-5} mm Hg. These numbers are deceptive, however, because the pressure differences between outside and inside are comparable ($760 - 10 = 750$ mm Hg in the first instance, compared with $760 - 10^{-5} = 760$ mm/Hg in the second instance). Therefore, *any* evacuated container must be regarded as an implosion hazard.

Vacuum distillation apparatus often provides some of its own protection in the form of heating mantles, column insulation, and the like; however, this is not sufficient because an implosion would scatter hot, flammable liquid. An explosion shield and a face mask should be used to protect the worker (see Sections I.F.1 and 4).

Equipment at reduced pressure is especially prone to rapid changes in pressure. This can create large pressure differences within the apparatus that can push liquids into unwanted locations, sometimes with very undesirable consequences (see Section I.G.2).

Water, solvents, or corrosive gases should not be allowed to be drawn into a building vacuum system. When the potential for such a problem exists, a water aspirator should be used as the vacuum source.

Mechanical vacuum pumps should be protected by using cold traps, and their exhausts should be vented to an exhaust hood or to the outside of the building (see Section I.G.2). If solvents or corrosive substances are inadvertently drawn into the pump, the oil should be changed before any further use. The belts and pulleys on such pumps should be covered with guards.

Glass Vessels

Glass vessels at reduced pressure are capable of collapsing violently, either spontaneously (if cracked or in some other way weakened) or from an accidental blow. Therefore, pressure and vacuum operations should be conducted behind adequate shielding (see Section I.F.4). It is advisable to check for flaws such as star cracks, scratches, or etching marks each time vacuum apparatus is used. Only round-bottomed or thick-walled, flat-bottomed flasks (e.g., Pyrex) specifically designed for operations at reduced pressure should be used as reaction vessels. Repaired glassware is subject to thermal shock and should be avoided, whenever possible, for operations at reduced pressure.

Dewar Flasks

These flasks are capable of collapsing as a result of thermal shock or a very slight scratch by a stirring rod. They should be shielded, either by a layer of friction tape or by enclosure in a wooden or metal container, to reduce the hazard of flying glass in case of collapse.

Desiccators

Glass vacuum desiccators should be made of Pyrex or similar glass. They should be completely enclosed in a shield or wrapped with friction tape in a grid pattern that leaves the contents visible and at the same time guards against flying glass should the vessel collapse. Various plastic (e.g., polycarbonate) desiccators now on the market reduce the implosion hazard and may be preferable.

Cold Traps

Cold traps should be of sufficient size and low enough temperature to collect all condensable vapors present in a vacuum system and should be interposed between the system and the vacuum pump. They should be checked frequently to guard against their becoming plugged by the freezing of material collected in them.

The common practice of using acetone-dry ice as a coolant should be avoided. Isopropanol or ethanol work as well as acetone and are cheaper, less toxic, less flammable, and less prone to foam on addition of small particles of dry ice. Dry ice and liquefied gases used in refrigerant baths should always be open to the atmosphere; they should never be used in closed systems where they could develop uncontrolled and dangerously high pressures.

After completion of an operation in which a cold trap has been used, the system should be vented. This venting is important because volatile substances that have collected in the trap may vaporize when the coolant has evaporated and cause a pressure buildup that could blow the apparatus apart. In addition, the oil from some pumps can be sucked back into the system.

Extreme caution should be exercised in using liquid nitrogen as a coolant for a cold trap. If such a system is opened while the cooling bath is still in contact with the trap, oxygen may condense from the atmosphere, which, if the trap contains organic material, will create a highly explosive mixture. Thus, a system that is connected to a liquid nitrogen trap should not be opened to the atmosphere until the trap has been removed. Also, if the system is closed after even a brief exposure to the atmosphere, some oxygen (or argon) may have already condensed. Then, when the liquid nitrogen bath is removed or when it evaporates, the condensed gases will vaporize with attendant pressure buildup and potential blowup.

Assembly of Vacuum Apparatus

Vacuum apparatus should be assembled so as to avoid strain. This requires that joints be assembled in a way that allows various sections of the apparatus to be moved if necessary without transmitting strain to the necks of the flasks. Heavy apparatus should be supported from below as well as by the neck.

Vacuum apparatus should be placed well onto the bench or into the hood where it will not be easily struck by passers by or the hood doors.

I.D.4 SAFE HANDLING OF CERTAIN VENDOR-SUPPLIED COMPRESSED GASES

ACETYLENE

Acetylene (C_2H_2) is special in several ways. First, it has the highest positive free energy of formation of any compound that most people ever encounter and is, hence, the most thermodynamically unstable common substance. Second, it has an exceedingly wide explosive range (from 2 to 80%). Furthermore, in addition to being explosive and shock sensitive, C_2H_2 can deflagrate in the absence of air to give (presumably) oligomers or polymers. Its stability is markedly enhanced by the presence of small amounts of other compounds, such as methane and, in fact, C_2H_2 from cylinders is rather safe to handle, because it is dissolved in acetone. For some uses, such as the preparation of acetylides, it is necessary to scrub the gas, thus removing the acetone. Such purified C_2H_2 is incomparably more dangerous than acetylene straight from the tank. Among other peculiarities of pure acetylene is the fact that its stability seems to be related to the diameter of the pipe used to transport it; it is actually less stable in wide-bore piping. The handling and use of C_2H_2 under pressure is extremely hazardous. In the absence of compelling reasons to the contrary, all reactions and operations involving C_2H_2 should be run in a pressure laboratory that has the necessary facilities, as well as expertise and experience, for its safe handling.

Handling Procedures

Under no circumstances should C_2H_2 be used under pressure in unbarricaded equipment. In any case, the lowest pressure necessary for the desired work should always be used.

A pressure of 103 kPa (15 lbf/in.^{2}) is generally accepted as presenting the maximum allowable hazard for supply lines and regulator systems. However, even below this pressure, a serious hazard still exists, particularly in closed systems containing more than 1 liter of gaseous C_2H_2.

Equipment

Only Underwriters Laboratory-approved equipment should be used in acetylene service. Acetylene forms shock-sensitive and explosive compounds, including copper and silver acetylides. Alloys of these metals, including solders, should not be used for C_2H_2 service unless they have been specifically approved for this purpose. If it is known or suspected that acetylides have been formed, a supervisor or safety officer should be consulted for safe methods of disposal (see Chapter II.E).

Contaminated piping should not be used in C_2H_2 service. Acetylene reacts with explosive violence with oxidizing agents such as chlorine or oxygen. Explosive decomposition is known to be initiated by a variety of conditions, particularly elevated temperatures. When an acetylene cylinder is connected to a pressure reactor, a valving system should be used to prevent flashback into the supply system.

Only pressure regulators approved for C_2H_2 service should be used. These are fitted with a flame arrestor. All repairs or modifications of C_2H_2 regulators should be done by qualified personnel.

Used gages may be contaminated and should be put in C_2H_2 service only after thorough reconditioning and inspection by qualified personnel. Only gages that have Bourdon tubes constructed of stainless steel or an alloy containing less than 60% copper should be used for acetylene service. Ordinary gages usually contain brass and bronze parts that can lead to acetylide formation.

Cylinders of C_2H_2 must be protected from mechanical shock. The C_2H_2 is in solution, under pressure, in a porous, acetone-impregnated, monolithic filler and is safe to handle only in this state. Such cylinders must always be vertically positioned and have the valve end up when gas is withdrawn from them.

Purification

The purification of C_2H_2 at atmospheric pressure and room temperature is most efficiently done by scrubbing the gas through concentrated sulfuric acid and caustic traps. Activated alumina (F.1 grade), an all-purpose solid absorbent, is recommended where purification over a solid is desired. Activated carbon should be avoided because the heat of absorption may be sufficient to trigger thermal decomposition of the C_2H_2.

Anhydrous Ammonia

A direct flame or steam jet must never be applied against a cylinder of ammonia (NH_3). If it becomes necessary to increase the pressure in a cylinder to promote more rapid discharge, the cylinder should be moved into a warm room. Extreme care should be exercised to prevent the temperature from rising above 50°C.

Only steel valves and fittings should be used on NH_3 containers. Neither copper, silver, nor zinc nor their alloys should be permitted to come into contact with ammonia.

Respiratory protective equipment of a type approved by the U.S. Mine Safety and Health Administration (MSHA) and NIOSH for anhydrous NH_3 service should always be readily available in a place where this material is used and so located as to be easily reached in case of need. Proper protection should be afforded the eyes by the use of goggles or large-lens spectacles to eliminate the possibility of liquid NH_3 coming in contact with the eyes and causing injury.

Leaks may be detected with the aid of sulfur tapers or sensitive papers. Both of these items and instructions as to their use may be procured from the cylinder supplier.

Boron Trichloride

Toxicity

The fumes of boron trichloride (BCl_3) are irritating to the eyes and mucous membranes. The material is, however, generally described as much less toxic than chlorine.

Handling Procedures

Although BCl_3 has a very low pressure at normal temperatures, it should be handled with care. The use of goggles and gloves is recommended (see Chapter III.B).

A trap should always be used when transferring the liquid or gas into a solution to prevent impurities from being sucked back into the cylinder, which might create an explosive mixture.

A BCl_3 cylinder contains a large amount of liquid and should never be permitted to become overheated. If the container is completely filled with liquid, an increase in temperature could build up tremendous pressure and cause the container to burst. For this reason, closed containers such as cylinders should be filled only by experts in the field.

Boron trichloride cylinders and valves should be equipped with a safety device containing a fusible metal that melts at approximately 70°C.

Boron Trifluoride

Toxicity

At high concentrations, boron trifluoride (BF_3) causes burns to the skin similar to those caused by hydrogen fluoride, although BF_3 burns do not penetrate as deeply as do hydrogen fluoride burns (see Section I.B.7).

Handling Procedures

Boron trifluoride has one characteristic that occasionally proves to be quite disconcerting. In contact with the atmosphere, the gas forms dense white fumes. Even after a cylinder valve has been tightly closed, the fumes will linger around the outlet for as much as 0.5 hour. This frequently causes the user to believe that the valve itself is leaking. In addition, the gas is inherently difficult to control through valves and piping, and even the best of equipment is apt to show slight signs of leaking, which will make an abundance of fumes.

It is essential when using BF_3 to have a trap in the delivery tube to prevent impurities from being sucked back into the cylinder. Certain chemicals, if drawn back into a BF_3 cylinder, can build up tremendous pressure that may cause the cylinder to burst.

Every BF_3 valve is equipped with a device consisting of a platinum disc in back of a plug containing a metal that will melt at approximately 70°C. Frequently, a similar safety device is inserted in the base of the cylinder.

Carbon Monoxide

Toxicity

Carbon monoxide (CO) is a direct and cumulative poison. It combines with the hemoglobin of the blood to form a relatively stable compound, carboxyhemoglobin, and renders it useless as an oxygen carrier. When about one-third of the hemoglobin has entered into such combination, the victim dies. The gas is a treacherous poison because of its odorless character and insidious action. Exposure to 1500-2000 ppm CO in air for 1 hour is dangerous, and exposure to 4000 ppm is fatal in less than 1 hour. Headache and dizziness are the usual symptoms of CO poisoning, but occasionally the first evidence of poisoning is the collapse of the patient.

Handling Procedures

Carbon monoxide should be used only in areas that have adequate ventilation at all times. A trap or vacuum break should always be used to prevent impurities from being sucked back into the cylinder.

CHLORINE

Toxicity See Safety Data Sheet (Section I.E.2).

Handling Procedures

Respiratory protective equipment of a type approved by MSHA and NIOSH for chlorine (Cl_2) service should always be readily available in places where this substance is used and so located as to be easily reached in case of need. Proper protection should be afforded the eyes by use of goggles or large-lens spectacles to eliminate the possibility of liquid Cl_2 coming in contact with the eyes and causing injury. Chlorine should be used only by experienced or properly instructed persons (see also Section I.E.2).

Only auxiliary valves and gages designed solely for Cl_2 use should be used. Stainless steel equipment should not be used.

Every precaution should be taken to avoid drawing liquids back into Cl_2 containers as their contents are being exhausted. To avoid this possibility, the valve should be closed immediately after the container has been emptied.

Chlorine leaks may be detected by passing a rag dampened with aqueous ammonia over the suspected valve or fitting. White fumes indicate escaping chlorine gas.

CHLOROMETHANE (METHYL CHLORIDE)

Toxicity

Methyl chloride (CH_3Cl) has a slight, not unpleasant odor, that is not irritating and may pass unnoticed unless a warning agent is added. Exposure to excessive concentrations of CH_3Cl is indicated by symptoms similar to those of alcohol intoxication—drowsiness, mental confusion, nausea, and possibly vomiting.

Handling Procedures

Because methyl chloride may, under certain conditions, react with aluminum or magnesium to form materials that ignite or fume spontaneously in contact with air, contact with these metals should be avoided.

Suitable respiratory protective equipment (see Section I.F.4) should be available and kept in proper working order. Cannister-type gas masks should have a fresh charge of suitable absorbent in the cannister. These masks, however, are unsafe at concentrations higher than 50 times the TLV; at these concentrations, a positive-pressure air-line respirator must be used and, at concentrations immediately dangerous to life and health (IDLH), a pressure-demand self-contained breathing apparatus or a positive-pressure air-line respirator that has escape-cylinder provisions must be used.

Chlorine Trifluoride

Toxicity

Liquid chlorine trifluoride (ClF_3) causes deep penetrating burns on contact with the body. The effect may be delayed and progressive, as in the case of burns by hydrogen fluoride (see Section I.B.7). The hazard of exposure to ClF_3 in the atmosphere is at least as great as that of chlorine.

Other Properties

Chlorine trifluoride reacts vigorously with water and most oxidizable substances at room temperature, frequently with immediate ignition. It reacts with most metals and metal oxides at elevated temperatures. In addition, it reacts with silicon-containing compounds and can, thus, support the continued combustion of glass, asbestos, and such. Chlorine trifluoride forms explosive mixtures with water vapor, ammonia, hydrogen, and most organic vapors. The substance resembles elemental fluorine in many of its chemical properties, and procedures for handling it, as well as precautions against accident in use, are closely parallel.

Handling Procedures

Because of the extreme activity of ClF_3, the area in the vicinity of a ClF_3 cylinder and its associated apparatus should be well ventilated and cleared of easily combustible material.

Chlorine trifluoride should always be removed from the cylinder as a gas. Cylinders of it should never be connected directly to apparatus containing a liquid absorbing medium. A trap should be inserted in the line between the cylinder and the reaction vessel adequate to contain the entire volume of liquid that might be drawn into the ClF_3 line.

Protective measures against ClF_3 are not fully developed, and entry into zones contaminated with the substance should be avoided. Only air-line or oxygen-supplied respiratory protective equipment (see Section I.F.4) is advised for personal protection against atmospheres contaminated with ClF_3. Gas masks, rubber gloves, rubber aprons, and face shields give protection only at low concentrations of gas and, if brought into local contact with a ClF_3 leak, may inflame. Thorough flushing with an inert gas should precede any opening of apparatus that has contained ClF_3.

FLUORINE

Toxicity

Fluorine (F_2) causes deep penetrating burns on contact with the body, an effect that may be delayed and progressive, as in the case of burns by hydrogen fluoride (see Section I.B.7). The hazard of exposure to F_2 in the atmosphere is at least as great as that of chlorine.

Other Properties

Fluorine reacts vigorously with most oxidizable substances at room temperature, frequently with immediate ignition, and with most metals at elevated temperatures. In addition, it reacts vigorously with silicon-containing compounds and can thus support the continued combustion of glass, asbestos, and such. Fluorine forms explosive mixtures with water vapor, ammonia, hydrogen, and most organic vapors.

Handling Procedures

Because of the high activity of F_2, the area in the vicinity of a F_2 cylinder and its associated apparatus should be well ventilated and cleared of easily combustible material.

When a cylinder of F_2 is to be opened, the user should be protected by a suitable shield (see Section I.C.2) and the valve should be opened by remote control. Any apparatus that is to contain F_2 under pressure should be surrounded by a protective barrier. Fluorine cylinder valves are not adapted to fine adjustment, and the flow of F_2 from a cylinder should therefore be controlled by a needle valve located close to the cylinder and operated by remote control.

All equipment that may be in contact with F_2 should be completely dry.

Protective measures against F_2 are not fully developed, and entry into zones contaminated with this substance should be avoided. Only positive-pressure atmosphere-supplying respiratory protective equipment (see Section I.F.1) is advised and, if IDLH concentrations are reached, a pressure-demand self-contained breathing apparatus or a positive-pressure air-line respirator that has escape-cylinder provisions must be used. Gauntlet-type rubber gloves (see Table 3), rubber aprons, and face shields give only temporary protection against F_2 and, if brought into local contact with a F_2 leak, may inflame. A thorough flushing of F_2 lines with an inert gas should precede any opening of the lines for any reason.

The reaction of F_2 with some metals is slow and results in the formation of a protective metallic fluoride film. Brass, iron, and aluminum and copper, as well as certain of their alloys, react in this way at standard temperatures and atmospheric pressure. Thus, these metals can be passivated by passing F_2 gas highly diluted with argon, neon, or nitrogen through tubing or over the surface with proper precautions and gradually increasing the concentration of F_2. As long as the protective coating is not cracked or dislodged to create a "hot spot," passivated apparatus is safe to use.

Once a fire has started that involves F_2 as an oxidizer, there is no effective way of stopping it other than shutting off the source of F_2. The area should be cleared and the fire allowed to burn itself out. No attempts should be made to extinguish the fire by using water or chemicals, as these act as additional fuel. Anyone in the vicinity of such a fire should wear impervious clothing and supplied-air or self-contained breathing apparatus.

HYDROGEN

Hydrogen (H_2) has an unusual and relatively unknown characteristic that, under certain conditions, presents a great hazard. Unlike other gases, the temperature of H_2 increases when the gas is expanded at a temperature higher than its inversion point (−80°C). This is known as the "inverse Joule-Thomson effect." It is well known that a cylinder of H_2 will sometimes emit a flash of fire when the cylinder valve is opened suddenly, permitting a rapid escape of gas. It is thought by some that the "inverse Joule-Thomson effect" plus the static charge generated by the escaping gas may cause its ignition.

Hydrogen has an extremely wide flammability range, the highest burning velocity of any gas and, although its ignition temperature is reasonably high, a very low ignition energy. Because no carbon is present, H_2 burns with a nonluminous flame that is often invisible in daylight. Hydrogen is nontoxic.

Handling Procedures

Hydrogen presents both combustion-explosion and fire hazards when released from containment. However, although its wide range of flammability and high burning rate accentuate these hazards, its low ignition energy, low heat of combustion on a volume basis, and nonluminous (low thermal-radiation level) flame are counteracting effects.

Because of its low ignition energy, when gaseous H_2 is released at high pressure, nominally rather small sources of heat (e.g., friction and static generation) often result in prompt ignition. Accordingly, H_2 is frequently thought of as "self-igniting" under these circumstances. When H_2 is released at low pressures, however, self-ignition is unlikely. Hydrogen combustion explosions are characterized by very rapid pressure increases that are extremely difficult to vent effectively. Open-air or space explosions have occurred from large releases of gaseous H_2.

Because of its very low boiling point, contact between liquid H_2 and air can result in condensation of air (see Section I.D.3) and its oxygen and nitrogen components. A mixture of H_2 and liquid oxygen is potentially explosive even though the quantities involved are likely to be small. Accidents from this source have been generally restricted to small containers of liquid H_2 that are handled open to the atmosphere and inside liquefaction equipment.

At ordinary temperatures, H_2 is very light, weighing only about 1/15 as much as air. The accordingly high diffusion rate makes it difficult for H_2 to accumulate in conventional structures and tends to reduce its combustion-explosion hazard.

Escaping gaseous H_2 seldom presents a "no fire" emergency situation because it either is ignited promptly or rises in the atmosphere rapidly. Hydrogen gas vaporizing from the cryogenic liquid near its normal boiling point is slightly heavier than air at 20°C and, this causes it, together with the visible fog of condensed water vapor created, to spread along the ground for sizable distances (depending on leak size and meteorological conditions). Because of the low gas density of vapors produced from vaporizing cryogenic liquid H_2, impounding or diked areas are not required.

Ignitable mixtures can extend well beyond the visible cloud. Such escapes can be controlled by water spray. Contact between water and pooled H_2 should be avoided to prevent increased vaporization unless the vapor can be controlled.

Water should be applied to containers of H_2 exposed to fire, and the flow of gas should be stopped if possible. Because a H_2 flame is often invisible in daylight and produces such low levels of thermal radiation, such flames have actually been walked into. Thus, great care should be exercised when approaching a H_2 fire.

HYDROGEN BROMIDE AND HYDROGEN CHLORIDE

Toxicity

Both hydrogen bromide (HBr) and hydrogen chloride (HCl) are highly toxic gases, being severely irritating to the upper respiratory tract. The acids formed neutralize the alkali of the tissues and can cause death as a result of edema or spasm of the larynx and inflamation of the upper respiratory system. Concentrations of 0.13-0.2% are lethal for human beings in exposures lasting a few minutes. However, because of their odor (see below), these gases provide adequate warning for prompt voluntary withdrawal from contaminated atmospheres.

These gases are also corrosive to the skin and mucous membranes and can cause severe burns. Exposure to high concentrations may also result in dermatitis. Contact with the eyes rapidly causes severe irritation of the eyes and eyelids.

Other Properties

Hydrogen bromide and hydrogen chloride are corrosive gases that have pungent, irritating odors. Although both are colorless, they fume in moist air because of their high solubility in water. In the cylinder under pressure, both exist in the form of a gas over a liquid [under such conditions, the cylinder pressure is equal to the vapor pressure of the substance contained—at 25°C, this is 4.22 MPa (613 lbf/in.^{2}) for HCl and 2.20 MPa (320 lbf/in.^{2}) for HBr.]. As long as liquid is present in the cylinder, the pressure will remain fairly constant. Although neither HBr nor HCl is combustible, both react with common metals to form hydrogen, which may form explosive mixtures with air.

Handling Procedures

Laboratory workers who handle HBr or HCl should wear protective apparel, including rubber gloves, suitable gas-tight chemical safety goggles, and clothing such as a rubber or plastic apron (see Sections I.F.2 and 3). Proper respiratory equipment should be available (see Section I.F.4).

These gases should be handled only in adequately ventilated areas. A check valve, vacuum break, or trap should always be used to prevent foreign materials from being sucked back into the cylinder because this can cause the development of dangerous pressures.

Leaks of HBr or HCl will be evident by the formation of dense white fumes on contact with the atmosphere. Small leaks of HCl can be detected by holding an open bottle of concentrated ammonium hydroxide near the site of the suspected leak; the formation of dense white fumes confirms the existence of a leak. Cylinder-valve leaks can usually be corrected by tightening the value packing nut (by turning it clockwise as viewed from above) (see Chapter II.E).

Hydrogen Sulfide

Toxicity See Safety Data Sheet (Section I.E.2).

Handling Procedures

Respiratory protective equipment (see Section I.F.4) approved by MSHA for hydrogen sulfide (H_2S) service should always be readily available in places where this material is used and so located as to be easily reached in case of need. A gas mask should be used only when the concentration of H_2S present is low.

Hydrogen sulfide should never be used from a cylinder without reducing the pressure through a suitable regulator attached directly to the cylinder.

Nickel Carbonyl

Toxicity

Nickel carbonyl [$Ni(CO)_4$] is poisonous when taken into the lungs in the gaseous form. The effect of the gas is that of poisoning by finely divided nickel over the moist pulmonary surfaces, from which it is absorbed in soluble form and spreads throughout the system. In nonfatal cases, it is gradually eliminated by normal body processes. Symptoms of carbonyl poisoning are giddiness, a sick feeling (sometimes with vomiting), and short or painful breathing.

Handling Procedures

Nickel carbonyl is both toxic and flammable. No one should be permitted to work with the material unless fully familiar with it. In direct sunlight, both the liquid and the gas will flash.

Although $Ni(CO)_4$ has an odor sometimes described as being like "brick dust," little confidence should be placed in detection of the odor because some people are unable to recognize it.

Excellent ventilation, either through hoods, fans, or strong drafts, is of utmost importance when $Ni(CO)_4$ is used (see Section I.B.8).

Nitric Oxide

Nitric oxide (NO) does not exist as such in significant concentrations in atmospheric air but, in the presence of oxygen, is transformed into nitrogen dioxide, a very toxic substance.

Nitrogen Dioxide

Toxicity

Nitrogen dioxide (NO_2) is classed as a primary irritant, acting primarily on the lungs and to a lesser extent on the upper respiratory tract. It is certainly one of the most insidious of the gases. The inflammation of the lungs may cause only slight pain, but the edema that results may easily cause death. One hundred ppm of NO_2 in air is a dangerous concentration for even a short exposure, and 200 ppm may be fatal in a short time. Nitrogen dioxide gas is reddish brown, has an irritating odor, and must be avoided by the use of an air-purifying respirator equipped with an acid-gas cartridge or canister; at concentrations greater than 50 times the TLV, a positive-pressure atmosphere-supplying respirator must be used and, in IDLH atmospheres, a pressure-demand self-contained breathing apparatus or a positive-pressure air-line respirator that has escape-cylinder provisions is required.

Handling Procedures

Nitrogen dioxide is a deadly poison, and no one should work with a cylinder of this substance unless they are fully familiar with its handling and its toxic effect. Ventilation is extremely important, and respiratory protective equipment should always be available. Only stainless steel fittings should be used.

Oxygen

Oils, greases, and other readily combustible substances should never come in contact with oxygen cylinders, valves, regulators, gages, and fittings. Oil and oxygen (O_2) may combine with explosive violence. Therefore, valves, regulators, gages, and fittings used in O_2 service must not be lubricated with oil or any other combustible substance. Oxygen cylinders or apparatus should not be handled with oily hands or gloves.

Oxygen regulators, hoses, and other appliances should not be interchanged with similar equipment intended for use with other gases. Cylinders of O_2 should not be stored near flammable materials, especially oils, greases, or any substance likely to cause or accelerate fire. Oxygen is not flammable, but supports combustion. Once a pure-oxygen fire begins, almost anything, including metal, will burn.

Phosgene

Toxicity See Safety Data Sheet (Section I.E.2).

Handling Procedures

Phosgene ($COCl_2$) is a deadly poison, and no one should work with this substance unless they are fully familiar with proper handling procedures and its toxic effect. Ventilation is extremely important, and respiratory protective equipment should be available (see Section I.F.4). Corrosion problems are not serious, and brass fittings may be used.

In case of a leak in a $COCl_2$ cylinder, the brass cylinder cap should be affixed as tightly as possible and the cylinder placed in the coolest spot that is available. The manufacturer should be notified at once (see Section II.E.6).

Sulfur Dioxide

Respiratory protective equipment (see Section I.F.4) of a type approved by MSHA for sulfur dioxide (SO_2) service should always be readily available in a place where this substance is used and so located as to be easily reached in case of need. When liquid SO_2 is used, the eyes should be protected by the use of goggles or large-lens spectacles (see Section I.F.1).

Leaks of SO_2 may be detected by passing a rag dampened with aqueous ammonia over the suspected valve or fitting. White fumes indicate escaping sulfur dioxide gas.

I.E

Known Hazards of and Specific Precautions for a Selected Group of Laboratory Chemicals

I.E.1 GENERAL CONSIDERATIONS

The following pages present a compilation of the chemical, physical, and physiological properties of 33 substances.

The number chosen for inclusion is arbitrary; the list is intended to be exemplary rather than exhaustive, for several reasons. For many of the hundreds of thousands of known chemicals, few, if any, toxicological data are available. Space limitations preclude inclusion of detailed information for more than a small fraction of the chemicals for which data are available. Moreover, the rapid progress of chemical and toxicological research ensure that new substances will be created every day and that new information on toxicology will have become available since these pages were assembled.

Each substance on the list satisfies some or all of the following criteria:

1. it is in common laboratory use;
2. under some conditions, it can constitute a known hazard; and
3. its properties and hazards may be encountered among a variety of chemicals and sometimes in many members of its class of chemicals.

In many cases, the description includes a cross-reference to another section of this report for instructions on proper handling of the substance.

It should be evident that the absence of a substance from this list *must not be interpreted as an indication that it is harmless.* On the contrary, the laboratory worker is well advised to treat any chemical with respect and to adhere to good safety practice as outlined elsewhere in these guidelines.

COMBINED EFFECTS OF CHEMICALS

The user of these sheets should keep in mind that they contain information on pure substances and that, because of the innumerable possible combinations of chemicals, it is not feasible to describe all of the conceivable circumstances in which a new hazard can be created. Additional hazard may exist because of the formation of new products or by-products, because of impurities, or because of synergistic effects. The product of the reaction of ammonia and iodine, two common normally nonexplosive chemicals, is the highly shock-sensitive explosive, nitrogen triiodide. Syntheses of halogenated phenols may give rise to polyhalogenated dibenzofurans or polyhalogenated dibenzo-*p*-dioxins as unavoidable, although limited, by-products. Benzene is sometimes present in toluene, and β-naphthylamine is present as an impurity in α-naphthylamine; often, an impurity may be present in such small concentrations that, despite its toxic properties, it presents no significant hazard and this is usually true of these examples. But it should be kept in mind that impurities do present a potential hazard and that the actual hazard posed by an impurity, dependent on its concentration, must be judged in each specific case. The effect of ethanol enhances the hepatotoxic effect of carbon tetrachloride (see Safety Data Sheet).

Because the nature of the materials present in a given reaction mixture may not be known, especially in the research laboratory where new preparations are constantly being developed, it is wise to maintain at least the same level of safe practice in the disposal of chemical wastes and residues as in the actual preparative procedure. This is especially important in the case of the nonvolatile residues from distillations, in which impurities may be present in concentrated form.

Odor as an Indicator

The threshold detection limits for odors of chemicals given in these lists are necessarily approximate because of the wide variation in the sensitivity of individuals to specific odors; an obvious corollary is that the absence of odor is not a reliable guide to safe concentration levels in the laboratory environment. The hazard of a particular substance depends on its physical properties and toxicity and how it is being used. The most hazardous type of substance is one that is volatile and highly toxic but has only a faint odor or causes olfactory fatigue. For example, hydrogen cyanide has poor warning properties and hydrogen sulfide rapidly produces olfactory fatigue.

Hazardous Properties of Classes of Chemicals

Individual chemicals of a class, e.g., aliphatic amides, alcohols, and such, vary so widely in their hazardous properties that it is not possible to generalize for the class. The difference in toxicity between ethanol and methanol by the oral route exemplifies this point. Nevertheless, it is certainly true that many members of a class of compounds may have common or related physical and toxicological properties.

Therefore, prudence suggests that, until contrary information becomes available, it is advisable to assume that a hazard of a known substance may also be characteristic of its new close relative. For example, mercury, bismuth, osmium, lead, and other heavy metals frequently form toxic derivatives. Many diazo compounds and azides are explosives, and several polycyclic aromatic substances are carcinogens. Organophosphorus compounds may be neurotoxins.

I.E.2 SAFETY DATA SHEETS

The terms used in the Safety Data Sheets are defined below.

CARCINOGENIC Causing malignant (cancerous) tumors (OSHA, NIOSH, and FDA consider any tumor to be either a cancer or a precursor of a cancer)

ONCOGENIC Causing tumors

TUMORIGENIC Causing tumors

MUTAGENIC Causing a heritable change in the gene structure

EMBRYOTOXIC Poisonous to an embryo (without necessarily poisoning the mother)

TERATOGENIC Producing a malformation of the embryo

HUMAN CARCINOGEN A substance that has been shown by valid, statistically significant epidemiological evidence to be carcinogenic to humans

EXPERIMENTAL CARCINOGEN A substance that has been shown by valid, statistically significant experimental evidence to induce cancer in animals

ACGIH American Conference of Governmental Industrial Hygienists

NIOSH National Institute for Occupational Safety and Health

LC_{50} The concentration in air that causes death of 50% of the test animals: The test animal and the test conditions should be specified; the value is expressed in mg/liter, mg/m^3, or ppm

LD_{50} The quantity of material that when ingested, injected, or applied to the skin as a single dose will cause death of 50% of the test animals: The test conditions should be specified; the value is expressed in g/kg or mg/kg of body weight

ALC The approximate lethal concentration in air for experimental animals: The test animal and the test condition should be specified; the value is expressed in mg/liter, mg/m^3, or ppm

TLV®-TWA The threshold limit value established by the ACGIH: The time-weighted average concentration for a normal 8-hour workday or 40-hour workweek to which nearly all workers may be repeatedly exposed, day after day, without adverse effect

PEL Permissible exposure limits for the workplace, set by regulation and enforced by OSHA; most of these limit values were originally set, by consensus, by the ACHIH to assist industrial hygienists in implementing exposure control programs. As law, these are listed in 29 CFR 1910.1000 and subject to revision through the regulatory process.

CFR Code of Federal Regulations

ACETYL PEROXIDE

CAS Registry No.: 110-22-5

Synonyms: Peroxide, diacetyl

Structure

$$CH_3-\overset{\overset{\displaystyle O}{\|}}{C}-O-O-\overset{\overset{\displaystyle O}{\|}}{C}-CH_3$$

Physical Properties

Molecular Weight: 118.1
Physical Form: Solid or colorless crystals or liquid
Melting Point: 30°C
Boiling Point: 63°C (21 mm Hg)
Explosion Hazard: Severe hazard when shocked or exposed to heat
Solubility: Slightly soluble in water; soluble in alcohol and hot ether

Toxicity and Hazard

Acetyl peroxide is irritating to the eyes, skin, and mucous membranes via the oral and inhalation routes. Application of two drops of a 30% solution (in dimethyl phthalate) has caused very severe corneal damage to rabbits.

Acetyl peroxide is a powerful oxidizing agent and can cause ignition of organic materials on contact. There are reports of detonation of the pure material; the 25% solution also has explosive potential, and inadvertent partial evaporation of even weak solutions can create explosive solutions or shock-sensitive crystalline material.

Special Handling Provisions

Acetyl peroxide is nearly always stored and handled as a 25% solution in an inert solvent. A face shield and rubber gloves should be worn when handling the substance, and a safety shield or hood door should be in front of apparatus containing it.

ACROLEIN

CAS Registry No.: 107-02-8

Synonyms: 2-Propenal, acrylaldehyde

Structure

$$CH_2{=}CHC(=O){-}H;\ C_3H_4O$$

Physical Properties

Molecular Weight: 56.06
Physical Form: Colorless to yellowish liquid
Melting Point: −87°C
Boiling Point: 52.7°C
Flash Point: Less than −18°C
Specific Gravity (liquid): 0.8427 (20°C/20°C)
Vapor Density (air = 1): 1.94
Vapor Pressure: 214 mm Hg (20°C)
Solubility: 20.8% by weight in water (with which it forms an azeotrope boiling at 52.4°C containing 97.4% acrolein)
Odor: Pungent and intensely irritating; threshold = 0.3-0.4 ppm

Toxicity and Hazard

After 1 min exposure to 1 ppm acrolein, volunteers have experienced slight nasal irritation; moderate nasal irritation and almost intolerable eye irritation with lacrimation developed in 5 min. At an acrolein level of 5 ppm, the latter effects were seen in 1 min. Inhalation of air containing 10 ppm acrolein may be fatal in a few minutes. Inhalation sufficient to cause intense lacrymation and nasal irritation may lead to slowly developing pulmonary edema in the course of 24 hours. Liquid acrolein in the eye or on the skin can produce serious injury.

Acrolein is highly toxic to animals via the inhalation and the oral routes: ALC (rats, 4-hour inhalation) = 9 ppm; LD_{50} (rats, oral) = 46 mg/kg. It is moderately toxic via skin absorption: LD_{50} (rabbits) = 200 mg/kg. Liquid acrolein is corrosive to skin, and skin sensitization can also occur. A 1% solution of acrolein caused severe eye injury to rabbits, and acrolein vapors can also cause damage to the eyes.

Ninety-day continuous exposure to acrolein at 0.21 and 0.23 ppm had no adverse effects on rats, guinea pigs, monkeys, or dogs. There was also no effect when the same species were exposed to 0.7 ppm for 6 weeks; however, similar exposures at 3.7 ppm produced toxic effects in the monkeys and dogs.

Special Handling Provisions

The TLV and the OSHA PEL for acrolein are 0.1 ppm (0.25 mg/m^3) as an 8-hour time-weighted average. The exposure limit over any 15-min period is 0.3 ppm.

Although the irritating odor of acrolein provides a useful warning, its high toxicity makes it advisable to carry out laboratory operations using it in a hood. Because of its high volatility and flammability, acrolein should not be handled near open flames.

ACRYLONITRILE

CAS Registry No.: 107-13-1

Synonyms: AN, cyanoethylene, 2-propenenitrile, VCN, vinyl cyanide

Structure

$CH_2=CHC\equiv N$; C_3H_3N

Physical Properties

Molecular Weight: 53.06
Physical Form: Colorless, mobile liquid
Melting Point: −82°C
Boiling Point: 77°C
Flash Point: −1°C
Autoignition Temperature: 481°C
Explosive Limits: 3.1-17% by volume in air
Specific Gravity: 0.806 (20°C/4°C)
Vapor Density (air = 1): 1.83
Vapor Pressure: 80 mm Hg (20°C)

Toxicity and Hazard

Depending on the amount and rapidity of absorption into the body, acrylonitrile can produce nausea, vomiting, headache, sneezing, weakness, light-headedness, asphyxia, and even death by inhalation, skin contact, or inadvertent ingestion. These toxic effects may be partially due to conversion of acrylonitrile to cyanide in the body.

The liquid can irritate eye and skin, and blistering has resulted after prolonged, apparently harmless, contact with previously contaminated clothing. The previously established workroom air concentration of 20 ppm appears adequate to prevent most adverse health effects of acrylonitrile. OSHA reduced the workplace exposure limit to 2 ppm with a ceiling value of 10 ppm for a single daily 15-min excursion following reports of cancer induction in animals and concern regarding its possible carcinogenicity to humans. Exposures at or lower than these levels should afford protection against that health risk.

Acrylonitrile is a flammable liquid. Fire hazard exists when this compound is exposed to heat, flame, or oxidizing agents. The substance also presents a moderate explosion hazard when exposed to flame. Acrylonitrile can react violently with strong acids, amines, strong alkalis, or bromine.

Special Handling Provisions

Acrylonitrile is regulated as a human carcinogen by OSHA (29 CFR 1910.1045). The PEL is 2 ppm as an 8-hour time-weighted average or 10 ppm as averaged over any 15-min period. Dermal or eye contact with liquid acrylonitrile is also prohibited. Where feasible, worker exposure must be controlled by engineering methods or work practices. Laboratory hoods that have been demonstrated to provide sufficient protection should be used, and closed systems are recommended for laboratory operations. Use of gloves (see Table 3) and goggles when handling liquid acrylonitrile is also recommended. OSHA regulations also require that exposure monitoring be conducted for all acrylonitrile operations to determine the airborne exposure levels for workers. In situations where the 15-min or 8-hour exposure limits are exceeded and engineering or administrative controls are not feasible, respiratory protection must be employed based on the expected exposure level. In cases of unknown concentration or fire fighting, supplied-air or self-contained breathing apparatus with a full facepiece operated in the positive-pressure mode is required. There are other detailed requirements in the OSHA standard related to housekeeping, waste disposal, hygiene facilities, employee training, and medical monitoring. Managers and laboratory supervisors should review these requirements before starting work with acrylonitrile.

ANILINE

CAS Registry No.: 62-53-3

Synonyms: Aminobenzene, benzenamine, phenylamine

Structure

$-NH_2$

$C_6H_5NH_2$

Physical Properties

Molecular Weight: 93.12
Physical Form: Colorless, oily liquid
Melting Point: −6.3°C
Boiling Point: 184°C
Flash Point: 158°C (closed cup)
Specific Gravity (liquid): 1.0217 (20°C)
Vapor Density (air = 1): 3.22
Vapor Pressure: 0.67 mm Hg (25°C)
1.0 mm Hg (30.6°C)
10.0 mm Hg (68.3°C)
Solubility: 3.5 g/100 ml water (20°C); miscible in most organic solvents
Odor: Characteristic odor and burning taste; threshold = 0.5 ppm

Toxicity and Hazard

An oral lethal dose of aniline for humans is 15-30 ml. Skin contact is the most common route of entry; the outstanding feature of aniline poisoning in humans is cyanosis due to formation of methemoglobin. The symptoms of severe exposure are cyanosis, headache, weakness, dizziness, nausea, and chills. Onset of symptoms, however, may be delayed up to 4 hours.

Aniline is moderately toxic via the skin and inhalation routes: LD_{50} (rabbits; skin) = 1540 mg/kg, ALC (rats, 4-hour inhalation) = 550 ppm. It is slightly toxic to animals via the oral route: LD_{50} (rats) = 633 mg/kg. Aniline is a moderate eye irritant and can cause mild skin irritation.

Rats fed 10, 30, or 100 (mg/kg of body weight)/day of aniline hydrochloride for 1 year experienced decreased red blood cell counts and hemoglobin concentrations and alterations of the spleen. Dogs, rats, mice, and guinea pigs exposed to 5 ppm aniline in the air daily for 6 months showed only slight methemoglobin. Exposure to 20 or 35 ppm caused 18-43% mortality; changes were seen in the liver, kidney, and spleen, and there was a marked effect on the composition of the blood.

Lifetime feeding studies in animals thus far do not answer the question of carcinogenicity but, if aniline is a carcinogen, it is not a potent one in animals. Animal studies now in progress should answer the question. Epidemiological studies of aniline workers have not found a relationship between aniline exposure and bladder tumors.

Special Handling Provisions

The TLV for aniline is 2 ppm (10 mg/m^3) as an 8-hour time-weighted average. The OSHA PEL is 5 ppm. These limits include a warning about the potential contribution of skin absorption to the overall exposure.

Because aniline, like many aromatic amines, is a rather toxic substance that readily penetrates the skin, it should be handled carefully. Most laboratory operations should be carried out in a hood, and skin contact should be avoided by appropriate use of protective apparel, e.g., rubber gloves and aprons.

BENZENE

CAS Registry No.: 71-43-2

Synonyms: Benzol, phenyl hydride, coal naphtha, mineral naphtha

Structure

C_6H_6

Physical Properties

Molecular Weight: 78.1
Physical Form: Colorless liquid
Melting Point: 6°C
Boiling Point: 80°C
Flash Point: −11°C
Explosive Limits: 1.3-7.1% by volume in air
Specific Gravity: 0.8787 (20°C/4°C)
Vapor Density (air = 1): 2.8
Vapor Pressure: 95 mm Hg (20°C)
Solubility: Miscible with most organic solvents

Toxicity and Hazard

In humans, acute inhalation exposure to benzene can produce a picture of acute delirium, characterized by excitement, euphoria, and restlessness and, if the exposure is significantly high, the symptoms may progress to depression, drowsiness, stupor, and even unconsciousness. The concentration required to produce this symptom complex is 1000 ppm or higher. These concentrations will also produce irritation of the eyes, nose, and respiratory tract.

Chronic inhalation exposure to 25-50 ppm of benzene can produce changes in the blood picture that include macrocytosis, decrease in the total red blood count, decrease in platelets, decrease in the hemoglobin concentration, or decrease in leukocytes. Any or all of these hematologic effects may be seen in any individual. Usually, the worker will be asymptomatic while these effects are observed in the blood picture. Continued exposure at somewhat higher concentrations (probably more than 100 ppm) can insidiously result in a more severe blood picture that includes leukopenia or even aplastic anemia, with symptoms of headaches, dizziness, loss of appetite, nervousness, irritability, and perhaps bleeding manifestations, i.e., nosebleeds, easy bruisability, or hematuria. Severe cases may have fatal outcomes.

Recently, a number of reports have been published that describe leukemia in workers who have had aplastic anemia. These cases have been reported in Italy and Turkey in workers exposed to grossly high concentrations of benzene. In addition, there is some indication that an excess of leukemia may occur without a preceding picture of aplastic anemia in workers who have been repeatedly exposed to benzene at concentrations of more than 100 ppm.

Special Handling Provisions

The current OSHA PEL for benzene [29 CFR 1910.1000(b) (Table Z-2)] is 10 ppm as an 8-hour time-weighted average, 25 ppm for a ceiling concentration for time periods such that the 8-hour TWA is not exceeded, and a peak above the ceiling of 50 ppm for no more than 10 min.

Benzene is a flammable liquid and should not be exposed to heat or flame. An explosion hazard also exists when its vapors are exposed to flame. Benzene may react vigorously with oxidizing agents such as bromine pentafluoride, chlorine, chromic acid, nitryl perchlorate, oxygen, ozone, perchlorates, aluminum chloride plus fluorine perchlorate, sulfuric acid plus permanganates, potassium peroxide, silver perchlorate plus acetic acid, and sodium peroxide.

Benzene operations in laboratories should be carried out in closed systems or in laboratory hoods that have been shown to have adequate protection factors to prevent significant worker exposure. When contact with liquid benzene is possible, skin-protection measures (see Table 3) should be employed. Before starting work with benzene, the worker should consult the OSHA standard, which requires more stringent precautions than does Procedure A (see Section I.B.8).

BENZO[*a*]PYRENE

CAS Registry No.: 50-32-8

Synonyms: 3,4-Benzpyrene

Structure

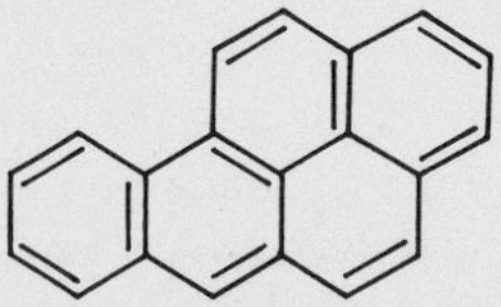

$C_{20}H_{12}$

Physical Properties

Molecular Weight: 252.3

Physical Form: Yellowish plates
Melting Point: 179°C
Boiling Point: 311°C
Solubility: Soluble in toluene, benzene, and xylene; sparingly soluble in methanol and ethanol

Toxicity and Hazard

There is very little information available on the acute (single-dose) toxicity of benzo[*a*]pyrene. Its acute oral toxicity is very low, probably because it is poorly absorbed by the gastrointestinal tract. Single contact with a 1% solution in toluene does not cause skin irritation, but repeated contact can cause systemic effects.

The TLV committee of ACGIH has rated benzo[*a*]pyrene as an occupational substance "suspect of oncogenic potential for workers." Although there are insufficient data to prove its carcinogenicity in humans, benzo[*a*]pyrene is a well-established animal carcinogen, affecting a variety of tissues including lungs, skin, and stomach. It is known to occur in coal-tar and other carcinogenic mixtures and has been identified as an active constituent of carcinogenic pitch.

Special Handling and Provisions

No TLV for benzo[*a*]pyrene has been set. Its carcinogenic potency in animals is high enough to justify the use of Procedure A (see Section I.B.8) when handling more than a few milligrams in the laboratory.

BIS(CHLOROMETHYL)ETHER

CAS Registry No.: 542-88-1

Synonyms: Methane, oxybis(chloro)-; BCME; chloromethyl ether; dichloromethyl ether; ether, bis(chloromethyl)

Structure

$ClCH_2OCH_2Cl$; $C_2H_4Cl_2O$

Physical Properties

Molecular Weight: 115.00
Physical Form: Colorless liquid
Melting Point: −41.5°C

Boiling Point: 104°C
Specific Gravity (liquid): 1.315 (20°C/4°C)
Vapor Density (air = 1): 4.0
Solubility: Miscible in all proportions with ethanol, ether, and many organic solvents; decomposes in water to give HCl and formaldehyde.
Odor: Suffocating

Toxicity and Hazard

Because of the high volatility of bis(chloromethyl)ether (BCME), inhalation is the route of exposure that presents the greatest hazard to humans. BCME vapor is severely irritating to the skin and mucous membranes and can cause corneal damage that heals slowly. The substance has caused lung cancer in humans.

BCME is highly toxic to animals via inhalation: LD_{50} (rats, 7-hour inhalation) = 7 ppm. It is moderately toxic via the oral and skin routes: LD_{50} (rats, oral) = 280 mg/kg; LD_{50} (rabbits, skin) = 368 mg/kg. Its vapors are strongly irritant to the eyes of rats.

Rats and hamsters subjected to 10 or 30, 6-hour exposures of 1 ppm BCME showed evidence of tracheal and bronchial hyperplasia, as well as effects on the central nervous system.

BCME is carcinogenic to mice following inhalation, skin application, or subcutaneous administration. In newborn mice, it is carcinogenic after a single subcutaneous exposure. In the rat, it is carcinogenic by inhalation and subcutaneous administration. BCME is a lung carcinogen in humans.

Special Handling Provisions

The TLV for BCME is 0.001 ppm (1 ppb; 5 $\mu g/m^3$). The substance is classified by ACGIH as a human carcinogen. OSHA has classified BCME as a cancer-suspect agent and has stringent regulations (29 CFR 1910.1008) for its use if its concentration in a material exceeds 0.1%. The regulations, which call for more precautions than does Procedure A (see Section I.B.8), should be consulted before starting work with BCME.

BROMINE

CAS Registry No.: 7726-95-6

Structure

Br_2

Physical Properties

Molecular Weight: 159.83
Physical Form: Dark reddish-brown liquid
Melting Point: −7.27°C
Boiling Point: 58.8°C
Flash Point: None
Specific Gravity (liquid): 3.11 (20°C/4°C)
Vapor Density (air = 1): 3.5
Vapor Pressure: 175 mm Hg (20°C)
Solubility: Soluble in alcohol, ether, chloroform, and carbon disulfide
Odor: Irritating and penetrating; threshold = 1.5-3.5 ppm.

Toxicity and Hazard

Fourteen mg/kg of Br_2 is a lethal oral dose for humans. Inhalation of Br_2 has caused coughing, nosebleeds, dizziness, and headache, followed after some hours by abdominal pain, diarrhea, and skin rashes. Severe irritation of the respiratory passages and pulmonary edema can also occur. Lacrimation occurs at levels of less than 1 ppm. It is reported that 40-60 ppm are dangerous for short exposures, and 1000 ppm can be fatal. The substance produces irritation and destruction of the skin with blister formation. Severely painful and destructive eye burns may result from contact with either liquid or concentrated vapors of Br_2.

Bromine is moderately toxic to animals via the inhalation route and slightly toxic via the oral route: LC_{50} (mice, 1.7-hour inhalation) = 240 ppm; LD_{50} (rats, oral) = 3100 mg/kg. The respiratory irritation threshold for Br_2 in rats is 1.4 ppm.

Rats fed 0.01 mg/kg Br_2 for 6 months experienced changes in their conditioned reflexes and several blood indexes. Rats, mice, and rabbits inhaling 0.2 ppm of Br_2 for 4 months developed disturbances in the functions of their respiratory, nervous, and endocrine systems; 0.02 ppm did not cause any adverse effects.

Special Handling Provisions

The TLV and the OSHA PEL for Br_2 are 0.1 ppm (0.7 mg/m^3) as an 8-hour time-weighted average. The exposure limit suggested by ACGIH is 0.3 ppm over any 15-min period.

Splash goggles and rubber gloves (see Table 3) should be worn when handling more than a few milliliters of pure liquid Br_2. Although the irritating odor of Br_2 provides a warning, it is best to carry out laboratory operations with it in a hood. Accidental contact with the skin must be immediately counteracted by washing with water. A worker whose clothing has been doused with liquid Br_2 is in severe danger unless the affected clothing is removed immediately.

CARBON DISULFIDE

CAS Registry No.: 75-15-0

Synonyms: Carbon bisulfide, dithiocarbonic anhydride, sulphocarbonic anhydride

Structure

S=C=S ; CS_2

Physical Properties

Molecular Weight: 76.14
Physical Form: Colorless liquid
Melting Point: −108.6°C
Boiling Point: 46.3°C
Flash Point: −30°C (closed cup)
Flammable Limits: 1.25-50.0% by volume in air
Autoignition Temperature: 100°C
Specific Gravity: 1.2626 (20°C)
Vapor Density (air = 1): 2.63
Vapor Pressure: 360 mm Hg (25°C)
Refractive Index: 1.6232 (25°C)
Conversions: 1 ppm = 3.11 mg/m^3
1 mg/m^3 = 0.32 ppm
Solubility: 0.22 g/100 ml water (22°C); miscible with alcohol, ether, and benzene
Odor: Almost odorless when pure; commercial samples may have a disagreeable odor due to trace of other sulfur compounds

Toxicity and Hazard

Carbon disulfide is rapidly absorbed when inhaled, and inhalation can produce acute poisoning. Poisoning can also occur from ingestion; death has been known to occur after ingestion of as little as 15 ml.

Acute poisoning by ingestion or inhalation can produce narcosis, accompanied by delirium. This may progress to areflexia, paralysis, coma, and death. Mental disorders or polyneuritis have been reported as sequelae. It is reported that exposure in excess of 500 ppm is required before acute effects will be noted.

Poisoning resulting from chronic exposure was reported frequently in the older literature but rarely in the last 10 to 20 years. Chronic exposure has resulted in neuropsychiatric manifestations with mental disorders, including psychoses, weakness, paralysis, parkinsonism, and blindness or in polyneuritis with pain along nerves, loss of strength, and paresthesias. In milder exposures, the reported effects have been attributed to cerebral vascular damage with symptoms related to central nervous system damage involving pyramidal, extrapyramidal, and pseudobulbar tracts. There are also reports of hypertension, renal damage, elevated cholesterol, and early arteriosclerosis. Some recovery from these effects is the rule, but such recovery is slow, occurring over months or years, and some paralysis may persist. There have also been reported effects on the reproductive system in both sexes, with women having menstrual disorders, abortions, and infertility, and males having spermatic disorders.

Carbon disulfide is a very flammable substance; a steam pipe or even a hot radiator can ignite its vapors. It also has a very wide range of explosive concentrations and thus should not be exposed to heat, flame, sparks, or friction. Carbon disulfide reacts violently with aluminum, chlorine, azides, cesium azide, chlorine monoxide, ethylene diamine, ethyleneimine, fluorine, lithium azide, nitric oxide, nitrogen tetroxide, sulfuric acid plus permanganates, potassium, potassium azide, rubidium azide, zinc, and various other oxidizing agents. When CS_2 is used to desorb organic materials from activated charcoal, as in the case of air sample analysis, a significant amount of heat can be liberated.

Special Handling Provisions

OSHA regulations require that worker 8-hour time-weighted average exposures not exceed 20 ppm; ceiling levels of 30 ppm are acceptable to the point that the 8-hour TWA is not exceeded, and additional peak exposures up to 100 ppm for no more than 30 min are allowable. ACGIH has decreased (1980) the 8-hour time-weighted average exposure to 10 ppm and has noted that skin absorption can significantly contribute to toxic effect.

Gloves (see Table 3) and protective apparel should be used when handling liquid CS_2. As much as possible, laboratory operations should be confined to a hood that has protection factors high enough to prevent significant exposure or to closed systems.

CARBON TETRACHLORIDE

CAS Registry No.: 56-23-5

Synonyms: Halon 104, perchloromethane, tetrachloromethane

Structure

CCl_4

Physical Properties

Molecular Weight: 153.8
Physical Form: Liquid
Melting Point: −23°C
Boiling Point: 76.5°C
Specific Gravity: 1.5940 (20°C/4°C)
Vapor Density (air = 1): 5.3
Vapor Pressure: 115 mm Hg (25°C)
Solubility: 0.080% in water (25°C); miscible with most organic solvents

Toxicity and Hazard

Although inhalation of CCl_4 can cause depression of the central nervous system with dizziness, headaches, depression, mental confusion, and even unconsciousness, such effects probably are the result of exposure at concentrations of 100-500 ppm, and serious poisonings rarely occur. Ingestion of as little as 4 ml of CCl_4 has been reported to be fatal. Many deaths have occurred from accidental ingestion; the early initial symptom is central nervous system depression, which usually clears the second day. Then, if the dose has been large enough, the victim may become jaundiced and, if the dose is sufficiently large, he or she may die in a few days. If the dose is smaller, the liver effects partially abate, only to have the victim go into renal failure with anuria, oliguria, uremia, proteinuria, and possibly death. Acute inhalation has produced almost the identical picture. In addition, occasionally, very brief inhalations of CCl_4 have been reported to produce sudden death thought to be due to ventricular fibrillation.

Chronic inhalation of CCl_4 at concentrations of 10-100 ppm has resulted in liver damage, which can be detected by abnormal liver function tests, and liver biopsies have disclosed centrilobular necrosis. Chronic exposure in animals has resulted in the appearance of cirrhosis and of hepatomas. There is little information regarding effects of chronic exposure in humans other than the laboratory abnormalities as described. A few cases of cirrhosis and hepatic cancer have been reported, but causal relationship to CCl_4 is difficult to confirm or deny. Prolonged exposure of the skin to the solvent can result in extreme dryness and fissuring, with redness and some secondary infection.

Ingestion of alcohol has been implicated repeatedly as predisposing the worker to increased effects of liver damage from CCl_4 exposure.

Although CCl_4 is nonflammable, on exposure to heat or flame it may decompose with the formation of phosgene. Severe reaction has been observed with allyl alcohol, aluminum, tetraethylaluminum, barium, benzoyl peroxide plus ethylene, beryllium, bromine trifluoride, calcium hypochlorite, diborane, ethylene, dimethyl formamide, disilane, fluorine, lithium, magnesium, liquid oxygen, plutonium, silver perchlorate plus hydrochloric acid, potassium tert-butoxide, sodium, potassium, tetrasilane, trisilane, uranium, zirconium, and burning wax.

Special Handling Provisions

The current OSHA PEL and TLV are 10 ppm as an 8-hour time-weighted average, 25 ppm as a ceiling for any period of time provided the 8-hour average is not exceeded, and 200 ppm for 5 min in a 4-hour period; in 1980, the ACGIH proposed a change to 5 ppm for an 8-hour time-weighted average and a ceiling exposure level of 20 ppm for up to 15 min on the basis that CCl_4 is suspected to have carcinogenic potential in humans. ACGIH also states that skin contact may account for a substantial part of toxic response.

Because the carcinogenic potency of CCl_4 is low, Procedure B (see Section I.B.7) provides adequate protection for laboratory operations in which it is used. All operations should be carried out in a hood, not only because of the carcinogenicity of the substance, but also because of its other toxic effects and its volatility. Nitrile rubber is the recommended material for gloves and other protective clothing.

CHLORINE

CAS Registry No.: 7782-50-5

Structure

Cl_2

Physical Properties

Molecular Weight: 70.91
Physical Form: Greenish-yellow gas
Melting Point: −101°C
Boiling Point: −34.1°C
Flash Point: Not flammable
Specific Gravity (liquid): 1.424 (15°C)
Vapor Density (air = 1): 2.4
Vapor Pressure: 400 mm Hg (25°C)
Solubility: 1.46% in water (25°C)
Odor: Penetrating and irritating; threshold = 0.3 ppm

Toxicity and Hazard

Humans can generally detect the odor of chlorine at about 0.3 ppm. Minimal irritation of the throat and nose are noticed at about 2.6 ppm and painful irritation at about 3.0 ppm; at a range of 2.6-41.0 ppm, a group of "trained industrial hygienists" noted "strong irritation." The subjective response to chlorine is less pronounced with prolonged exposure.

Experimentally determined responses to chlorine by humans are not very consistent. Throat irritation occurs at about 6.6-15 ppm. However, an exposure for medical purposes of a large number of humans to 5-7 ppm for 1 hour did not result in serious or long-term consequences. Exposure to about 17 ppm causes coughing, and levels as low as 10 ppm may cause lung edema.

Human exposure to 14-21 ppm for 30 min to 1 hour is regarded as dangerous and may, after a delay of 6 or more hours, result in death from anoxia due to serious pulmonary edema. For rats, the LC_{50} (1 hour) = 293.

Chronic effects on humans from long-term low-level exposures have not been well documented. Animal exposures have indicated that prolonged exposure to approximately 1.7 ppm for 1 hour per day may cause deterioration in the nutritional state, blood alteration, and decreased resistance to disease.

Special Handling Conditions

The TLV and the OSHA PEL are 1 ppm (3 mg/m^3) as a ceiling. NIOSH has recommended a ceiling limit of 0.5 ppm over any 15-min period. The ACGIH 15-min exposure limit is 3 ppm.

Chlorine should be kept away from easily oxidized materials. Chlorine reacts readily with many organic chemicals, sometimes with explosive violence. Because of the high toxicity of chlorine, laboratory operations using it should be carried out in a hood (see Section I.B.7) and appropriate gloves (see Table 3) should be worn.

CHLOROFORM

CAS Registry No.: 67-66-3

Synonym: Trichloromethane

Structure

$CHCl_3$

Physical Properties

Molecular Weight: 119.39
Physical Form: Colorless liquid
Melting Point: −63.5°C
Boiling Point: 61.26°C
Flash Point: None
Specific Gravity: 1.49845 (15°C)
Vapor Pressure: 100 mm Hg (10.4°C)
Vapor Density (air = 1): 4.12
Solubility: 0.74% in water (25°C); miscible with most organic solvents

Toxicity and Hazard

Inhalation exposure to $CHCl_3$ at concentrations greater than 1000 ppm can produce dizziness, nausea, and headache. At higher concentrations, there can be disorientation and delirium progressing to unconsciousness. Such high exposures can also produce liver and possibly kidney damage. It is believed that $CHCl_3$ can sensitize the heart to adrenaline, so it may cause cardiac arrhythmias and possibly death.

High concentrations of the vapor can produce conjunctivitis. Liquid $CHCl_3$ in the eye will produce a painful corneal injury that usually heals in several days.

Chronic exposure to $CHCl_3$ at concentrations of 100-200 ppm has been reported to produce enlarged livers. Continued contact with the skin can produce drying, fissuring, and inflammation.

In experimental studies, prolonged ingestion of high levels of $CHCl_3$ by mice resulted in liver cancers and by rats, kidney tumors.

Although the fire hazard of $CHCl_3$ is slight, exposure to heat or flame can result in generation of phosgene gas. $CHCl_3$ reacts violently with acetone plus a base, aluminum, disilane, lithium, magnesium, nitrogen tetroxide, potassium, perchloric acid plus phosphorus pentoxide, potassium hydroxide plus methanol, potassium *tert*-butoxide, sodium, sodium hydroxide plus methanol, sodium methylate, or sodium hydride.

Special Handling Conditions

The current OSHA PEL for $CHCl_3$ is 50 ppm as an 8-hour time-weighted average. This standard is also a ceiling level that should not be exceeded for any 15-min period. The ACGIH currently recommends that $CHCl_3$ be treated as a suspect human carcinogen and recommends an 8-hour time-weighted average exposure of 10 ppm.

Although $CHCl_3$ has caused tumors in animals, its potency is low. Hence, Procedure B (see Section I.B.7) provides adequate protection during laboratory operations with it. The high volatility of $CHCl_3$ emphasizes the importance of a hood for such operations. Polyvinyl alcohol gloves provide the best hand protection.

DIETHYL ETHER

CAS Registry No.: 60-29-7

Synonyms: Ethane, 1,1′-oxybis-; ethyl ether; ether

Structure

$CH_3CH_2OCH_2CH_3$; $C_4H_{10}O$

Physical Properties

Molecular Weight: 74.12
Physical Form: Colorless liquid
Melting Point: −116.3°C (stable crystals)
Boiling Point: 34.6°
Flash Point: −45° (closed cup)
Explosive Limits: 1.9-36.5% by volume in air
Autoignition Temperature: 180°C
Specific Gravity (liquid): 0.7146 (20°C/20°C)

Vapor Density (air = 1): 2.55
Vapor Pressure: 438.9 mm Hg (20°C)
Solubility: 8.43 wt-% in water (15°C); 6.05 wt-% in water (25°C)
Odor: Sweetish, pungent (characteristic); threshold = 0.2 ppm

Toxicity and Hazard

Repeated exposure of humans to diethyl ether via inhalation has caused loss of appetite, exhaustion, headache, and other symptoms. General anesthesia occurs at a concentration of 3.6-6.5% in air. Human subjects found diethyl ether irritating to the nose but not to the eyes or throat at a level of 200 ppm. Acute overexposure produces vomiting, paleness, irregular respiration, and low pulse rates and body temperatures. The human oral lethal dose for diethyl ether is about 420 mg/kg.

Diethyl ether is slightly toxic to animals via the oral route: LD_{50} (rats) = 1700 mg/kg. It is a mild skin irritant. Its absorption through the skin is not usually great enough to cause a deleterious effect. Diethyl ether can cause eye irritation but not any permanent damage. It has very low toxicity via inhalation: LC_{50} (mice, 3-hour inhalation) = 42,500 ppm.

Diethyl ether is hazardous in several ways. Diethyl ether readily forms explosive peroxides on exposure to air, sometimes leading to explosive residues when it is distilled.

Special Handling Conditions

The TLV and the OSHA PEL for diethyl ether are 400 ppm (1200 mg/m^3) as an 8-hour time-weighted average. The exposure limit is 500 ppm over any 15-min period.

Because of its high volatility and flammability, diethyl ether should be used in a hood that has a spark-proof mechanical system and be kept well away from flames or sparking devices. It should be stored in a cool place, preferably an explosion-proof refrigerator. Opened bottles of diethyl ether, even those containing an oxidation inhibitor such as BHT, should not be kept more than a few months to avoid the hazard of peroxide formation. Uninhibited ether, such as that specifically prepared for anesthesia use, should be handled with particular care.

DIMETHYLFORMAMIDE

CAS Registry No.: 68-12-2

Synonym: DMF

Structure

$$HC(=O)-N(CH_3)_2;\ C_3H_7NO$$

Physical Properties

Molecular Weight: 73.1
Physical Form: Colorless, mobile liquid
Melting Point: −61°C
Boiling Point: 152.8°C
Flash Point: 136°C
Autoignition Temperature: 445°C
Explosive Limits: 2.2-15.2% by volume in air at 100°C
Specific Gravity: 0.9445 (25°C/4°C)
Vapor Density (air = 1): 2.51
Vapor Pressure: 3.7 mm (25°C)

Toxicity

Human overexposure to DMF may produce gastrointestinal effects. Employees in a French textile plant using DMF developed digestive symptoms including burning, nausea, vomiting, and stomach cramps, probably from irritation of the digestive mucosa. The symptoms subsided with absence from work and reappeared when work was resumed.

In experimental studies, animal exposures indicate that dimethylformamide is only slightly toxic via the inhalation and the oral routes: LC_{50} (rats) = 5000 ppm and LD_{50} (rats, oral) = 3967 mg/kg. It is moderately toxic when placed on the skin: LD_{50} (rabbits, percutaneous) = 4720 mg/kg. When placed on the skin, it produced slight inflammation of the skin, and when placed in the eye, it produced corneal injury.

Animal experimentation also indicates that dimethylformamide is not a teratogen.

The fire hazard of dimethylformamide is only moderate when the substance is exposed to heat or flame. Materials for which contact should be avoided include inorganic nitrates, bromine, chromic acid, organic nitrates, phosphorus pentoxide, and tetraethylaluminum.

Special Handling Provisions

The current OSHA PEL for dimethylformamide is 10 ppm as an 8-hour time-weighted average. It is advised that significant toxicity can result from skin contact. Therefore, gloves or other protective apparel (preferably of butyl rubber) should be worn when handling liquid DMF. It should also be noted that use of DMF as a solvent for toxic materials that are not ordinarily absorbed may increase their skin contact hazard.

DIMETHYL SULFATE

CAS Registry No.: 77-78-1

Synonyms: DMS; dimethyl monosulfate; sulfuric acid, dimethyl ester

Structure

$$CH_3O\overset{\overset{\displaystyle O}{\|}}{\underset{\underset{\displaystyle O}{\|}}{S}}{-}OCH_3\,;\ C_2H_6O_4S$$

Physical Properties

Molecular Weight: 126.13
Physical Form: Colorless, waterlike liquid
Melting Point: −31.8°C
Boiling Point: 188.8°C
Flash Points: 116°C (open cup)
83°F (closed cup)
Specific Gravity (liquid): 1.328 (20°C)
Vapor Density (air = 1): 4.35
Vapor Pressure: 0.54 mm Hg (20°C)
Solubility: 2.8 g/100 ml of water (18°C); hydrolyzes to sulfuric acid and methanol in water; soluble in alcohol, ether, and benzene
Odor: Slight; not distinctive

Toxicity and Hazard

Many cases of DMS poisoning have been reported. The common initial symptoms are headache and giddiness, with burning of the eyes. The patient's condition may worsen, with painful eyes, nose and throat irritation, loss of voice, coughing, difficulty in breathing and swallowing, vomiting, and diarrhea possible. The onset of these symptoms may be delayed for up to 10 hours.

Skin contact causes blistering and necrosis, and DMS can be absorbed through the skin in sufficient quantity to cause systemic intoxication. In the worst cases, there is severe inflammation of the mucous membranes and pulmonary injury that may be fatal; several deaths have occurred. For example, exposure to 97 ppm for 10 min was fatal.

DMS is moderately toxic to animals via the oral route: LD_{50} (rats) = 440 mg/kg. Undiluted DMS produced moderate to severe irritation when applied to the skin of guinea pigs; 1% DMS produced mild irritation. DMS does not cause skin sensitization in animals. Undiluted DMS applied to rabbit eyes produced severe injury. Even a 1-hour exposure to 58 ppm has resulted in permanent eye damage in rats. During a 4-hour exposure, 30 ppm DMS killed five out of six rats, but 15 ppm was not lethal.

DMS has been shown to be carcinogenic in the rat by inhalation, subcutaneous injection, and following prenatal exposure. By inhalation, tumors developed in rats following 1 hour per day exposures to 10 ppm DMS for 130 days.

Special Handling Conditions

The TLV for DMS is 0.1 ppm (0.5 mg/m^3) as an 8-hour time-weighted average. DMS is classified as being suspect of carcinogenic potential in humans by the ACGIH. The OSHA PEL for DMS is 1.0 ppm. These limits include a warning of the potential contribution of skin absorption to the overall exposure.

Procedure A (see Section I.B.8) should be used when handing more than a few grams of DMS in view of its fairly high carcinogenic potency in rats by inhalation and its ability to penetrate the skin. It is particularly important to avoid skin contact by appropriate use of rubber gloves, a rubber apron, and other protective apparel and to avoid inhalation of even low concentrations of vapor by working in a hood.

DIOXANE

CAS Registry No.: 123-91-1

Synonyms: *p*-Dioxane; 1,4-diethylene dioxide; 1,4-dioxane; 1,4-dioxacyclohexane

Structure

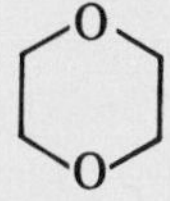

$C_4H_8O_2$

Physical Properties

Molecular Weight: 88.10
Physical Form: Colorless liquid
Melting Point: 11.8°C
Boiling Point: 101.3°C
Flash Point: 11°C
Specific Gravity (liquid): 1.03 (20°C/4°C)
Vapor Density (air = 1): 3
Vapor Pressure: 37 mm Hg (25°C)
Solubility: Miscible in all proportions with water, acetone, alcohols, and most organic solvents
Odor: Faintly alcoholic; threshold = 5.7 ppm.

Toxicity and Hazard

A worker who was exposed via the skin and inhalation routes to 500 ppm of dioxane for 1 week died. Autopsy revealed damage to the kidney, liver, and brain. Symptoms of inhalation overexposure include irritation of the upper respiratory tract, coughing, eye irritation, vertigo, headache, and vomiting. An airborne concentration of 300 ppm of dioxane causes irritation of the eyes, nose, and throat. At lower levels, exposure sufficient to cause harm can occur before one realizes it through smell or irritation. Prolonged or repeated skin contact can produce drying and fissuring of the skin. Dioxane forms explosive peroxides in contact with air, especially in the presence of moisture.

Dioxane is slightly toxic to animals via the skin: LD_{50} (rabbits) = 7600 mg/kg and has a very low toxicity via the oral route: LD_{50} (rats) = 7120 mg/kg. No skin irritation was seen in rabbits or guinea pigs from a 25% aqueous solution. However, dioxane is irritating to the skin on prolonged or repeated contact. Dioxane caused mild, transient injury in rabbit eyes. It is slightly toxic on inhalation. No serious symptoms were seen in guinea pigs exposed to 2000 ppm for several hours. However, exposure to 30,000 ppm for 3 hours was lethal.

Rats and mice given drinking water containing 5% dioxane for 30-60 days experienced severe liver and kidney damage. Animals exposed to 1000 ppm dioxane in air for 135 exposures also had liver and kidney damage. However, rats exposed to 111 ppm for 2 years experienced no significant adverse effects.

Dioxane is a weak animal carcinogen. Tumors developed in rats and guinea pigs at feeding levels of about 1% in the diet. Rats had nasal and liver tumors, and guinea pigs had gall bladder and liver tumors. In another study, rats receiving 0.1% dioxane in their water for their lifetime were without adverse effects. Rats exposed to 111 ppm of airborne dioxane for 2 years showed no compound-related effects.

Special Handling Provisions

Dioxane is the principal ingredient of Bray's solution (scintillation cocktail), and uninhibited solutions have been known to explode if left for a period of time.

The current TLV for dioxane is 50 ppm (180 mg/m^3) [with a notice (1980) of intended change to 25 ppm] as an 8-hour time-weighted average. The exposure limit is 100 ppm over any 15-min period. The OSHA PEL for dioxane is 100 ppm. These limits include a warning about the potential contribution of skin absorption to the overall exposure. NIOSH has recommended (1977) a ceiling of 1 ppm in any 30-min period.

Although dioxane has caused tumors in animals, these have occurred only by prolonged exposure to very large amounts and, hence, it is considered a carcinogen of such low potency that no special precautions beyond normal good laboratory practice are needed for working with it. Nitrile rubber is the preferred material for gloves and other apparel used to protect against skin contact. The high volatility of the compound requires that all laboratory operations with it be carried out in a hood. Because it is miscible in water, prompt washing is an effective way to remove dioxane that has accidentally contacted the skin.

ETHYLENE DIBROMIDE

CAS Registry No.: 106-93-4

Synonyms: 1,2-Dibromoethane; ethylene bromide; EDB

Structure

$BrCH_2CH_2Br$; $C_2H_4Br_2$

Physical Properties

Molecular Weight: 187.88
Physical Form: Colorless liquid
Melting Point: 9.79°C
Boiling Point: 132°C
Flash Point: Nonflammable
Specific Gravity (liquid): 2.1792 (20°C)
Vapor Density (air = 1): 6.5
Vapor Pressure: 12 mm Hg (25°C) 17.4 mm Hg (30°C)
Solubility: 0.43 g/100 ml water (30°C); miscible with most organic solvents
Odor: Sweet; threshold = 10 ppm

Toxicity and Hazard

The approximate oral lethal dose of EDB for humans is 5 ml. Skin adsorption of EDB can also cause death, and inhalation of the vapor can produce pulmonary edema. EDB can cause severe irritation to all exposed tissues, respiratory tract, skin, and eye. Systemic effects include central nervous system depression, kidney injury, and severe liver necrosis.

Ethylene dibromide is highly toxic to animals via inhalation. The maximum survival exposures of rats to EDB vapors in air are 3000 ppm for 6 min, 400 ppm for 30 min, and 200 ppm for 2 hours. It is moderately toxic via the oral and skin routes: LD_{50} (rats, oral) = 140 mg/kg; LD_{50} (rabbits, skin) = 300 mg/kg. EDB is markedly irritating to skin, and a 10% solution has caused serious but reversible corneal injury in rabbit eyes.

Rats were repeatedly exposed to 50 ppm EDB for 6 months. Half died from pneumonia and upper respiratory tract infections. Slight changes in the liver and kidney were seen.

EDB has induced a high incidence of tumors (squamous-cell carcinomas of the forestomach) in mice and rats following oral administration.

Special Handling Conditions

The 1979 TLV for EDB was 20 ppm (155 mg/m^3) as an 8-hour time-weighted average. The exposure limit is 30 ppm over any 15-min period. These limits included a warning about the potential contribution of skin absorption to the overall exposure. In 1980, the ACGIH put EDB in category A1b (human carcinogen). For this category, there is no assigned TLV, but the ACGIH recommends that those working with A1b carcinogens should be properly equipped to ensure virtually no contact with the carcinogen. The OSHA PEL for EDB is 20 ppm, and the acceptable maximum peak is 50 ppm for 5 min in any 8-hour time period.

On the basis of the carcinogenicity data for EDB, Procedure A (see Section I.B.8) should be followed when handling more than a few grams in the laboratory. Serious skin injury can occur from direct exposure to EDB. The substance can penetrate neoprene and several other types of plastic; therefore, gloves and other protective apparel of these materials provide only temporary protection if EDB spills on them.

FORMALDEHYDE

CAS Registry No.: 50-00-0

Synonyms: Formalin (this name applies to 30-55% aqueous solutions)

Structure

$$HC(=O)\text{–}H;\ CH_2O$$

Physical Properties

Molecular Weight: 30.03
Physical Form: Colorless gas or aqueous solution or solid polymer (paraformaldehye)
Melting Point: −92°C (gas)
Boiling Point: −19°C (gas); a 37% aqueous solution boils at about 98°C
Flash Points:
37% formaldehyde solution containing
6% methanol 72.2°C (closed cup)
10% methanol 63.8°C (closed cup)
15% methanol 50°C (closed cup)
Explosive limits: 7-73% by volume in air
Specific Gravity (liquid): 0.815 (20°C)
Vapor Density (air = 1): 1.075 (gas)
Vapor Pressure: 10 mm Hg (−88°C) (gas)
Solubility: Very soluble in water; soluble in ether, alcohol, and most organic solvents
Odor: Pungent and irritating; threshold = 1 ppm

Toxicity and Hazard

For humans, an oral dose of 90 ml of 37% formalin (about 520 mg formaldehyde per kg of body weight) is almost certainly fatal within 48 hours. However, 30 ml has been fatal and 120 ml nonfatal in certain cases. Inhalation of vapors may result in severe irritation and edema of the upper respiratory tract, burning and stinging of the eyes, and headache and has been known to cause death. Workers exposed to 2-10 ppm have experienced headaches, nausea, dizziness, and vomiting; lacrimation occurs at 4-5 ppm. For several minutes of exposure, 10 ppm or more is intolerable. Solutions of formaldehyde are irritating to the skin and can cause severe injury if splashed in the eye.

A study of formaldehyde-exposed workers showed an above-average incidence of chronic upper respiratory tract disease. Sensitization of the skin can result from repeated exposure.

Formaldehyde is moderately toxic to animals via inhalation: ALC (rats, 4-hour inhalation) = 250 ppm and is slightly toxic via the oral route: LD_{50} (rats) = 585 mg/kg. Either formaldehyde gas or formalin may cause skin irritation. Formaldehyde is a skin sensitizer. It is a severe eye irritant, causing delayed effects that are not appreciably eased by eye washing.

Rats receiving oral doses of 50, 100, or 150 mg/kg formaldehyde daily for 90 days showed no adverse effects except for a decrease in weight gain at the highest level. Dogs fed 100 mg/kg for 90 days showed a decrease in weight gain, but no other significant effects.

Preliminary data from a study that is still in progress have indicated the development of nasal cancers in rats exposed to 15 ppm formaldehyde for 18 months. In another study, central nervous system effects were seen among rats exposed to 0.8 ppm for 3 months, but not among rats at 0.03 ppm. Mice exposed to 41-163 ppm for up to 64 weeks showed no untoward effects.

Special Handling Conditions

The OSHA PEL for formaldehyde is 3 ppm (4.5 mg/m^3) as an 8-hour time-weighted average, with a 15-min ceiling of 5 ppm and a 10 ppm maximum peak. The TLV is a 2-ppm ceiling limit.

Laboratory operations with formalin in open vessels should be carried out in a hood or other local exhaust device (formaldehyde has such an objectionable odor that there may be little need for this admonition). Because repeated exposure to formaldehyde can lead to a formaldehyde allergy, it is well to avoid skin contact with aqueous solutions by appropriate use of neoprene, butyl rubber, or polyvinyl chloride gloves (see Table 3) and other protective apparel. Splash-proof goggles should be worn if there is any possibility of splashing formaldehyde in the eyes.

If the preliminary indications of carcinogenicity described above are confirmed during completion and assessment of the study, the use of procedure A or B (see Sections I.B.8 and 7, respectively) will be called for.

HYDRAZINE (and its salts)

CAS Registry No.: 302-01-2

Synonyms: None of significance; however, the following forms may be encountered: hydrazine hydrochloride, sulfate, etc.; hydrazine hydrate

Structure

H_2NNH_2 ; N_2H_4

Physical Properties

Molecular Weight: 32.0
Physical Form: Colorless, fuming, oily liquid
Melting Point: 2.0°C
Boiling Point: 113.5°C
Flash Point: 52°C
Explosive Limits: 4.7-100% by volume in air
Specific Gravity (liquid): 1.011 (15°C)
Vapor Density (air = 1): 1.11
Vapor Pressure: 10.4 mm Hg (20°C)
Solubility: Miscible with water and ethanol; insoluble in hydrocarbons
Odor: Ammonialike, fishy; threshold = 3-4 ppm

Toxicity and Hazard

Acute exposure to hydrazine vapors can cause respiratory tract irritation, excitement, convulsions, cyanosis, and a decrease in blood pressure. The liquid can severely burn the eye and skin. Hydrazine can cause fatty degeneration of the liver, nephritis, and hemolysis.

Hydrazine also poses a dangerous fire and explosion risk and can explode during distillation if traces of air are present.

Hydrazine is moderately toxic to animals via the inhalation, oral, and skin routes: LC_{50} (rats, 4-hour inhalation) = 570 ppm; LD_{50} (rats, oral) = 60 mg/kg; LD_{50} (rabbits, skin) = 283 mg/kg (hydrazine hydrate). It is a strong skin and mucous membrane irritant and a strong skin sensitizer. Hydrazine hydrate produced moderately severe irritation when applied to rabbit eyes.

After repeated oral, skin, or injection exposure, the effects noted include weight loss, weakness, vomiting, and convulsions. The chief histological finding is fatty degeneration of the liver. Among guinea pigs and dogs exposed to hydrazine in the air 5-47 times, the dogs showed liver damage, with lesser damage to the kidneys and lungs, while the guinea pigs had pneumonitis and partial lung collapse.

Hydrazine or hydrazine salts have been shown to be carcinogenic in mice after oral and intraperitoneal administration and in rats following oral dosing. By the oral route, effects were found at doses of 24-36 (mg/kg)/day in mice and 20 (mg/kg)/day in rats. No tumors were observed in Syrian golden hamsters after oral administration. The ACGIH has classified hydrazine as suspect of carcinogenic potential in humans.

Special Handling Conditions

The TLV for hydrazine is 0.1 ppm (0.1 mg/m^3) and the OSHA PEL is 1.0 ppm (1 mg/m^3) as 8-hour time-weighted averages. These limits include a warning about the potential contribution of skin absorption to the overall exposure. NIOSH has (1978) recommended a ceiling limit of 0.03 ppm in any 2-hour period.

When more than a few grams of hydrazine are to be used in the laboratory, Procedure A (see Section I.B.8) should be used because hydrazine is carcinogenic in animal tests, quite volatile, and readily absorbed through the skin. Moreover, it is a serious risk as an acute poison and a skin and eye irritant. Nitrile rubber is recommended for gloves and other protective apparel. Prompt washing with water effectively removes hydrazine from skin that it has splashed on.

Hydrazine should not be used in the vicinity of a flame or under conditions where sparks can occur, as an explosion or fire can result.

HYDROGEN CYANIDE

CAS Registry No.: 74-90-8

Synonyms: Hydrocyanic acid, HCN

Structure

HC≡N; CHN

Physical Properties

Molecular Weight: 27.03
Physical Form: Colorless liquid
Melting Point: −13.4°C
Boiling Point: 25.7°C

Flash Point: −17.8°C (closed cup)
Explosive Limits: 6-41% by volume in air
Specific Gravity (liquid):
 99.7% 0.690 (20°C/15.6°C)
 96% 0.703 (20°C/15.6°C)
Vapor Density (air = 1): 0.94
Vapor Pressure: 807 mm Hg (27.2°C)
Solubility: Miscible with water; soluble in alcohol and ether
Odor: Sweetish, characteristic; threshold = 2-5 ppm

Toxicity and Hazard

Hydrogen cyanide is among the most toxic and rapidly acting of all known substances. Exposure to high doses may be followed by almost instantaneous collapse, cessation of respiration, and death. At lower dosages, the early symptoms include weakness, headache, confusion, nausea, and vomiting. In humans, the approximate fatal dose is 40 mg via the oral route. Exposure to 3000 ppm HCN is immediately fatal, while 200-480 ppm can be fatal after 30 min. Exposure to 18-36 ppm HCN causes slight symptoms after several hours. The liquid is rapidly absorbed through the skin or the eyes.

Hydrogen cyanide is extremely toxic to animals via the oral and skin routes: oral LD_{50} (mice) = 3.7 mg/kg. However, it has little or no irritant effect on the skin. Hydrogen cyanide in the eye may cause some local irritation, which is of little significance because the attendant absorption may be hazardous to life. Hydrogen cyanide is highly toxic via inhalation: LC_{50} (rats, 5 min) = 503 ppm, and 100 ppm was fatal to rats in 1.5 hours.

No adverse effects were seen in rats fed HCN-fumigated food for 2 years. In this experiment, special feeding jars were designed to reduce volatilization of the HCN, and analysis of the two test diets 3 days after fumigation showed HCN levels of 100 and 300 ppm. Mice, rats, guinea pigs, rabbits, cats, dogs, pigeons, and a monkey were exposed to various concentrations of HCN. No adverse effects were observed in mice exposed to 40 ppm for 7 hours or in rats and mice exposed to 16 ppm for 16 hours. Concentrations of 31.5 ppm were generally safe for most species exposed for a few hours. Dogs subjected repeatedly to 30-min exposures of 45 ppm HCN exhibited cumulative effects, particularly central nervous system lesions, hemorrhages, and vasodilation.

Special Handling Conditions

The OSHA 8-hour PEL for HCN is 10 ppm with a 15-min TWA of 15 ppm. ACGIH has recently proposed that the basis of the 10 ppm TLV be changed from an 8-hour TWA to a ceiling concentration. These limits include a warning against the potential contribution of skin absorption to the overall exposure. In 1976, NIOSH recommended a limit of 4.7 ppm (5 mg/m^3 as CN) determined as a ceiling concentration, based on a 10-min sampling period.

Aside from its high toxicity, HCN has a low flash point and forms an explosive mixture with air over a wide range of concentrations. Moreover, traces of base can cause rapid spontaneous polymerization, sometimes resulting in detonation. Hence, HCN is very dangerous, anyone working with it should wear goggles and impervious gloves, and no one should work alone with it. In cases of overexposure to HCN, quick action is called for in removing the victim from the contaminated area, using amyl nitrate ampules to restore consciousness or, if breathing has stopped, artificial respiration. Medical assistance should be summoned as soon as possible, but the victim should not be left unattended. Speed in providing treatment is of the utmost importance.

(See also Section I.B.7.)

HYDROGEN SULFIDE

CAS Registry No.: 7783-06-4

Structure

HSH; H_2S

Physical Properties

Molecular Weight: 34.08
Physical Form: Colorless gas
Melting Point: −82.9°C
Boiling Point: −61.8°C
Explosive Limits: 4.3-46% by volume in air
Specific Gravity: 1.54
Vapor Density (air = 1): 1.189 (15°C)
Vapor Pressure: 8.77 atm (20°C)
Solubility: 437 ml/100 ml in water (0°C) and 186 ml/100 ml (40°C); also soluble in alcohol and petroleum solvents

Odor: "Rotten egg"; threshold = 0.2-0.003 ppm; odor appears sweet at 30 ppm and above; high concentrations deaden the sense of smell

Toxicity and Hazard

Hydrogen sulfide is extremely dangerous. Human exposure to relatively low concentrations of H_2S has caused corneal damage, headache, sleep disturbances, nausea, weight loss, and other symptoms suggestive of brain damage. Higher concentrations can cause irritation of the lungs and respiratory passages and even pulmonary edema. Exposure to 210 ppm for 20 min has caused unconsciousness, arm cramps, and low blood pressure. Coma may occur within seconds after one or two breaths at high concentrations and be followed rapidly by death. For example, workers exposed to 930 ppm H_2S for less than 1 min died.

Hydrogen sulfide is moderately toxic to animals via the inhalation route: LC_{50} (mice, 1 hour) = 673 ppm; LC_{50} (mice, 7.5 hours) = 140 ppm. Exposure to 10-13 ppm for 4-7 hours has caused eye irritation. Skin absorption of H_2S is slight and not considered significant. However, prolonged or repeated skin contact might cause mild irritation. Guinea pigs that had 0.78 $in.^2$ of their skin exposed to 100% H_2S vapors for 1 hour experienced slight swelling.

Rabbits exposed to 100 ppm H_2S for 30 min/day for 4 months showed changes in the blood (leucopenia and lymphocytosis). Animals continuously exposed for 90 days to 20 ppm exhibited pathologic changes of the lungs.

Special Handling Conditions

The TLV for H_2S is 10 ppm (14 mg/m^3) as an 8-hour time-weighted average. The short-term exposure limit (15 min) is 15 ppm. The OSHA PEL has a ceiling limit of 20 ppm and a peak of 50 ppm over any 10-min period. NIOSH (1977) has recommended a 10-min ceiling of 10 ppm.

Partly because of the disagreeable odor of H_2S, but also because of its toxicity, laboratory operations with it should be carried out in a hood. Cylinders of it should not be stored in small, unventilated rooms, as deaths have resulted from people entering such rooms containing a leaking cylinder.

(See Section I.B.7.)

METHANOL

CAS Registry No.: 67-56-1

Synonyms: Methyl alcohol, wood alcohol

Structure

CH_3OH; CH_4O

Physical Properties

Molecular Weight: 32.04
Physical Form: Colorless liquid
Freezing Point: −97.8°C
Boiling Point: 65°C
Flash Point: 12°C
Specific Gravity (liquid): 0.7915
Vapor Density (air = 1): 1.11
Vapor Pressure: 95 mm Hg (20°C)
Solubility: Completely miscible with water, ether, and most organic solvents
Odor: Mild, threshold = 3-8 ppm

Toxicity and Hazard

Methanol is well known to cause blindness in humans, but this usually results from drinking large quantities. Once absorbed, methanol is only very slowly eliminated. Severe exposure to the vapors of methanol can cause dizziness, central nervous system depression, shortness of breath, coma, and eventually death. Where the exposure is less severe, the first symptoms may be blurring of vision, photophobia, conjunctivitis, headache, gastrointestinal disturbances, dizziness, and a feeling of intoxication, followed by the development of definite eye lesions.

Methanol has a very low acute toxicity in animals via the inhalation and oral routes: ALC (rats, head only, 1-hour inhalation) > 145,000 ppm; LD_{50} (rats, oral) = 6200 mg/kg. The substance is moderately toxic via the skin: LD_{50} (rabbits) = 14,400 mg/kg. Methanol has caused moderate corneal opacity and conjunctival redness in rabbit eyes.

Methanol has shown a relatively low chronic toxicity in animal studies. For example, no adverse effects were seen in dogs exposed to 450-500 ppm of methanol in the air 8 hours daily for 379 days. Rats given 1% methanol in their drinking water for 6 months showed no significant effects. However, in another oral study in rats, toxic effects were found in the liver.

Special Handling Conditions

The TLV and the OSHA PEL for methanol are 200 ppm (260 mg/m^3) as an 8-hour time-weighted average. This limit includes a caution against skin contact. The exposure limit is 250 ppm over any 15-min period.

Although methanol is one of the safest solvents, it is best to carry out operations in a hood if significant amounts could escape into the laboratory atmosphere; for example, during recrystallization from boiling methanol in an open flask. If there are opportunities for significant hand contact, neoprene gloves should be worn. If a still safer solvent of similar properties seems called for, ethanol is often a good choice.

MORPHOLINE

CAS Registry No.: 110-91-8

Synonyms: Tetrahydro-1,4-oxazine; tetrahydro-1,4-isoxazine; diethyleneimide oxide

Structure

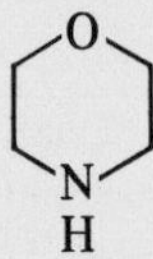

C_4H_9NO

Physical Properties

Molecular Weight: 87.12
Physical Form: Mobile, hygroscopic liquid
Melting Point: −4.9°C
Boiling Point: 128.9°C
Flash Point: 38°C (open cup)
Explosive Limits: Not defined
Specific Gravity: 1.002 (20°C)
Vapor Density (air = 1): 3.0
Vapor Pressure: 6.6 mm Hg (20°C)
Solubility: Miscible with water and most organic solvents

Toxicity and Hazard

Exposure to the vapors of morpholine can produce irritation of the skin and, if inhaled, the substance can cause irritation of the respiratory tract, similar to that produced by ammonium hydroxide. There can be a burning sensation of the nose and throat and coughing, and blurred vision can result from corneal edema. The liquid is a severe irritant to the skin and can produce ulceration of the conjunctiva.

In laboratory experiments, repeated exposure of rats at high concentrations (18,000 ppm) produced death, with damage to the lung, liver, and kidneys.

Morpholine is a moderate fire hazard when exposed to heat, flames, or oxidizing agents. Decomposition results in generation of oxides of nitrogen.

Special Handling Conditions

OSHA has established an allowable 8-hour time-weighted average exposure to morpholine of 20 ppm. It should be noted that skin contact can be a significant contributor to toxic effects. The ACGIH has established a TLV of 20 ppm for an 8-hour TWA and also a short-term exposure limit of 30 ppm for 15 min.

The TLV was established to prevent irritation of the respiratory tract and the effects on the eye.

NITROBENZENE

CAS Registry No.: 98-95-3

Synonyms: Nitrobenzol

Structure

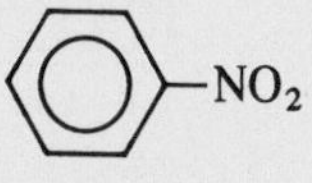

C_6H_5—NO_2

$C_6H_5NO_2$

Physical Properties

Molecular Weight: 123.11
Physical Form: Yellow, oily liquid

Melting Point: 5.7°C
Boiling Point: 211°C
Flash Point: 88°C (closed cup)
Specific Gravity (liquid): 1.19867 (24°C/4°C)
Vapor Density (air = 1): 4.24
Vapor Pressure: 10 mm (85.4°C); 100 mm (139.9°C)
Solubility: 0.2 g/100 ml water (20°C); miscible with most organic solvents
Odor: Oil of bitter almond; threshold = 0.5 ppm

Toxicity and Hazard

The oral lethal dose of nitrobenzene for humans is about 5 mg/kg. No immediate or delayed effects have been seen in humans after 30-60 min of exposure to 200-300 ppm of nitrobenzene. Exposure to 40-80 ppm for several hours, however, will cause slight symptoms. The symptoms of overexposure to nitrobenzene, whether by inhalation or by skin contact, are cyanosis due to methemoglobin formation in the blood, anoxia, weakness, and sometimes shock. Inflammation of the skin is sometimes seen. Absorption through the skin is the greatest hazard in the workplace. Onset of symptoms may be delayed for up to 4 hours.

Nitrobenzene is slightly toxic to animals via the oral route: LD_{50} (rats) = 640 mg/kg. It is moderately toxic via the skin: LD_{50} (rats) = 2100 mg/kg. It is a mild eye irritant.

Nitrobenzene transforms hemoglobin into methemoglobin on oral administration to rats. Rats exposed to about 0.01-0.02 ppm of nitrobenzene for 70-82 days experienced adverse central nervous system effects and inflammation of internal organs. A concentration of 0.0016 ppm in air caused no significant adverse effects in rats during 73 days of constant exposure.

Special Handling Conditions

The TLV and the OSHA PEL for nitrobenzene are 1 ppm (5 mg/m^3) as an 8-hour time-weighted average. These limits include a warning against the potential contribution of skin absorption to the overall exposure. The exposure limit is 2 ppm during any 15-min period.

Nitrobenzene, like most aromatic nitro compounds, readily penetrates the skin to cause the serious toxic effects described above. Hence, anyone using it in the laboratory should take care to avoid skin contact. If more than a few grams are being used, rubber gloves and other protective apparel may be needed. Although nitrobenzene is only moderately volatile, it is advisable to handle it in a hood.

PERACETIC ACID

CAS Registry No.: 79-21-0

Synonyms: Peroxyacetic acid, acetyl hydroperoxide

Structure

$$CH_3C(=O)\text{–}O\text{–}OH;\ C_2H_4O_3$$

Physical Properties

Molecular Weight: 76.05
Physical Form: Colorless liquid
Available Form: Most commonly, as 40% solution in acetic acid
Melting Point: 0.1°C
Freezing Point: Approximately −30°C
Boiling Point: 105°C
Flash Point: 105°C
Autoexplosion Temperature: 110°C

Toxicity and Hazard

Peracetic acid, like most peracids, is unstable. It has an acrid odor and is very irritating to skin, eyes, and upper respiratory tract. On the basis of animal test data, it is considered highly to moderately toxic by ingestion in single doses.

By oral administration to test animals, peracetic acid is moderately to highly toxic, causing severe irritation of the stomach and intestinal linings. It is moderately toxic by single dermal applications to rabbits. Repeated applications at relatively high dosages to the skin of mice produced skin irritation and indicated that peracetic acid is a potent tumor promotor and a weak carcinogen.

Special Handling Conditions

Because explosion is the greatest hazard of peracetic acid, the substance should be protected from sparks or physical shock and laboratory operations with it should be carried out behind a shield and in a hood that has explosion-proof equipment. Peracetic acid explodes at 100°C and decomposes at lower temperatures with the generation of oxygen. It reacts vigorously with organic materials.

Because of its high irritancy, care should be taken to prevent contact of peracetic acid with the skin, eyes, or upper respiratory tract. This is another reason why work with it should be carried out in a hood and also means that rubber gloves and a rubber apron should be worn as appropriate.

PHENOL

CAS Registry No.: 108-95-2

Structure

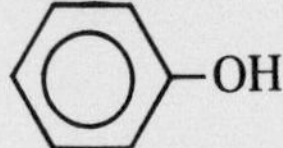

C_6H_5OH

Physical Properties

Molecular Weight: 94.11
Physical Form: Colorless to light pink solid
Melting Point: 41°C
Boiling Point: 182°C
Flash Point: 85°C (open cup)
Specific Gravity (liquid): 1.049 (50°C/4°C)
Vapor Density (air = 1): 3.24
Vapor Pressure: 0.3513 mm Hg (25°C)
Solubility: 6.7 g/100 ml water (16°C); miscible at 66°C; easily soluble in alcohol and other organic solvents
Odor: Characteristically sweet; threshold = 0.3 ppm

Toxicity and Hazard

In humans, lethal oral doses of phenol have ranged from 1 to 10 g. Severe phenol poisoning by ingestion is characterized by burns of the mouth and throat and rapid development of digestive disturbances, headache, fainting, vertigo, mental disturbances, collapse, and coma. Exposure to the vapor can produce marked irritation of the eyes, nose, and throat. Concentrated solutions are corrosive to the eyes and skin. Phenol readily penetrates the skin.

Phenol is moderately toxic to animals via the inhalation and the oral routes: LC_{50} (rats, 1-hour inhalation) = 312 ppm; LD_{50} (rats, oral) = 414 mg/kg. It is also moderately toxic via skin absorption: LD_{50} (rabbits) = 1120 mg/kg. Phenol is corrosive to the eyes and skin and may cause irritant dermatitis.

Guinea pigs were severely injured by inhalation of 25-50 of ppm phenol for 20 days. Damage was seen in the lungs, liver, kidneys, and heart.

Special Handling Conditions

The TLV and the OSHA PEL for phenol are 5 ppm (19 mg/m^3) as an 8-hour time-weighted average. The standard includes a warning about the potential contribution of skin absorption to the overall exposure. The exposure limit is 10 ppm over any 15-min period. NIOSH (1976) has recommended an 8-hour time-weighted average of 5.20 ppm with a limit of 15.6 ppm over any 15-min period.

Because phenol is a potent skin irritant, rubber gloves (see Table 3) should be worn when there is opportunity for significant skin contact.

PHOSGENE

CAS Registry No.: 77-44-5

Synonym; Carbonyl chloride

Structure

$$ClC(=O)-Cl;\ CCl_2O$$

Physical Properties

Molecular Weight: 98.92
Physical Form: Colorless gas
Melting Point: −104°C
Boiling Point: 8.3°C
Specific Gravity (liquid): 1.392 (19°C/4°C)
Vapor Density (air = 1): 3.4
Vapor Pressure: 563 mm Hg (0°C); 760 mm Hg (8.3°C); 1418 mm Hg (25°C)
Solubility: Decomposes in water with formation of CO_2 and HCl; soluble in most organic solvents and oils
Odor: Sweet (geraniumlike) at low levels, pungent and irritating at higher levels; threshold = 0.5-1 ppm

Toxicity and Hazard

In humans, the symptoms of overexposure to phosgene are dryness or a burning sensation in the throat, numbness, vomiting, and bronchitis. An airborne concentration of 5 ppm may cause eye irritation and coughing in a few minutes. The substance can cause severe lung injury in 1-2 min at a level of 20 ppm. Exposure to concentrations above 50 ppm is likely to be fatal.

Phosgene is extremely toxic to animals via inhalation. Thus, 74% of a group of rats died from exposure to 55-100 ppm for only 10 min. Liquid phosgene is likely to cause severe skin burns and eye irritation.

Pulmonary edema, bronchiolitis, and emphysema were found in cats and guinea pigs exposed to 2.5-6.25 ppm of phosgene/day for 2-41 days. A variety of animals exposed to 0.2 or 1.1 ppm for 5 hours per day for 5 days also had pulmonary edema.

Special Handling Conditions

The TLV and the OSHA PEL for phosgene are 0.1 ppm (0.4 mg/m^3) as an 8-hour time-weighted average. NIOSH has recommended a limit of 0.2 ppm over any 15-min period.

In the laboratory, work with phosgene should always be carried out within a hood. Unused quantities of phosgene greater than 1 g should be destroyed by reaction with water or dilute alkali.

(See Section I.B.7.)

PYRIDINE

CAS Registry No.: 110-86-1

Structure

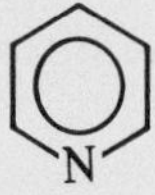

C_5H_5N

Physical Properties

Molecular Weight: 79.1
Physical Form: Colorless basic liquid
Melting Point: 42.1°C
Boiling Point: 115.6°C
Flash Point: 20.0°C (tag closed cup)
Explosive Limits: 1.8-12.5% by volume in air
Specific Gravity: 0.986 (15.5°C)
Vapor Density (air = 1): 2.73
Vapor Pressure: 20 mm Hg (25°C)

Solubility: Miscible with water and most organic solvents
Odor: Unpleasant; characteristic

Toxicity and Hazard

Acute exposure to pyridine can produce transient headaches, dizziness, or light-headedness. Insomnia, mental dullness, nausea, and anorexia have also been reported. Such effects have been reported in workers exposed to more than 100 ppm on a relatively regular basis. The liquid and vapor are irritating to the eyes, nose, and throat.

Chronic exposure has produced serious liver and kidney damage, with death having occurred once. This exposure was by ingestion, prescribed therapeutically, of about 2 ml/day. There are also reports of central nervous system effects. Continuing skin contact with the liquid can produce a dry, scaly, inflammation.

Special Handling Conditions

Pyridine presents a significant fire hazard when exposed to heat or flame. The danger of explosion is severe if the vapor contacts heat or flame. Pyridine reacts violently with chlorosulfonic acid, chromic acid, maleic anhydride, nitric acid, fuming sulfuric acid, perchromates, β-propiolactone, silver perchlorate, and sulfuric acid. On decomposition, cyanides can be liberated.

The current OSHA PEL for pyridine is 5 ppm as an 8-hour time-weighted average. Although the ACGIH recommends the same 8-hour exposure limit, it also recommends a short-term exposure limit of 10 ppm for 15 min. The odor threshold of approximately 1 ppm requires that pyridine be handled in a laboratory hood or with the use of some other local exhaust ventilation.

SODIUM AZIDE

CAS Registry No.: 26628-22-8

Structure

NaN_3

Physical Properties

Molecular Weight: 65.02
Physical Form: Colorless hexagonal crystals

Melting Point: Decomposes
Density: 1.846

Toxicity and Hazard

Sodium azide is highly toxic to humans. It is highly soluble in water and rapidly converted to hydrazoic acid, which may, therefore, be the ultimate toxic agent in a human exposed to sodium azide.

Acute exposure to hydrazoic acid vapor can produce irritation of the eyes, tracheal bronchitis, headache, possibly a dramatic decrease in blood pressure, weakness, pulmonary edema, and collapse.

Accidental ingestion of 50-60 mg of sodium azide has resulted in brief loss of consciousness, nausea, and severe headache, but recovery was rapid. In another incident, while acidifying 10 g of sodium azide, a chemist complained of dizziness, blurred vision, shortness of breath, and faintness following a few minutes of exposure. Hypotension and bradycardia were seen but, again, recovery was complete in 1 hour.

Acute poisoning in laboratory animals has established that sodium azide is highly toxic via the oral route: LD_{50}, rats = about 45 mg/kg. After a lethal dose, the animals showed respiratory distress and convulsions, followed by nervous system depression and death. After severe poisoning by the azide, rats exhibited lesions in the optic nerves and tracts.

Sodium azide has been used to control blood pressure therapeutically. Doses of 0.65-3.9 mg by mouth daily for up to 2.5 years lowered the blood pressure and reduced the transient pounding sensation of the head with no evidence of organic damage.

It is thought that a TLV of 0.2 mg/m^3 of sodium azide or 0.1 ppm of hydrazoic acid will prevent significant lowering of blood pressure or headache discomfort.

Special Handling Conditions

Sodium azide is one of the azides that is not explosive. It may, however, react violently with benzoyl chloride plus potassium hydroxide, bromine, carbon disulfide, chromium oxychloride, copper, lead, nitric acid, dimethylsulfate, and dibromomalononitrile. It is especially important that sodium azide not be allowed to come in contact with heavy metals (for example, by being poured into a lead or copper drain) or their salts; heavy metal azides detonate with notorious ease.

There is currently no OSHA PEL for sodium azide. The ACGIH recommends that exposure to this compound not exceed 0.1 ppm as an 8-hour time-weighted average. This level is also recommended as a ceiling that should not be exceeded even instantaneously during the day. Use of a laboratory hood that has a protection factor adequate to prevent significant worker exposure or a closed system is recommended for operations using sodium azide.

SODIUM CYANIDE

CAS Registry No.: 143-33-9

Structure

NaC≡N

Physical Properties

Molecular Weight: 49.02
Physical Form: White solid
Melting Point: 564°C
Boiling Point: 1496°C
Vapor Pressure: 1.0 mm Hg (817°C)
10.0 mm Hg (983°C)
Solubility: Readily soluble in water; slightly soluble in alcohol

Toxicity and Hazard

Sodium cyanide is among the fastest acting of all known poisons. The lethal oral dose for humans is 200 mg. The symptoms of cyanide overdose include weakness, headache, confusion and, occasionally, nausea and vomiting. Higher doses may be followed by almost instantaneous collapse, cessation of respiration, and immediate death. Solutions are irritating to the skin, nose, and eyes, and cyanide is adsorbed through the skin.

Sodium cyanide is highly toxic to animals via the oral route: LD_{50} (rats) = 6.4 mg/kg. It can be corrosive to the skin and the eyes, for it is highly alkaline. Sodium cyanide can also produce toxic symptoms via skin absorption and inhalation.

Special Handling Conditions

The TLV and the OSHA PEL for cyanide are both 5 mg/m^3 as an 8-hour time-weighted average. These limits include a warning of the potential contribution of skin absorption to the overall exposure. In 1976, NIOSH recommended that the 5-mg/m^3 limit be retained but that its basis be changed from an 8-hour TWA to a 10-min ceiling.

Dry cotton gloves should be worn when handling dry sodium cyanide.

Rubber gloves and splash-proof goggles should be worn when substantial amounts of sodium cyanide solution are used.

Acid must not be allowed to come in contact with sodium cyanide, as gaseous hydrogen cyanide will be liberated. (The special problems of hydrogen cyanide are discussed in Section I.B.7, along with safety and first aid measures that mostly apply to sodium cyanide as well.)

Tert-BUTYL HYDROPEROXIDE

CAS Registry No.: 75-91-2

Synonyms: Hydroperoxide, *tert*-butyl, cardox TBH

Structure

$(CH_3)_3C{-}O{-}OH$; $C_4H_{10}O_2$

Physical Properties

Molecular Weight: 90.12
Physical Form: Water-white liquid
Melting Point: −8°C
Boiling Point: 35°C
Flash Point: 27°C or above
Specific Gravity: 0.896 (20°C/4°C)
Vapor Density (air = 1): 2.07
Solubility: Moderately soluble in water; very soluble in organic (esters and alcohols) solvents and alkali metal hydroxide solutions

Toxicity and Hazard

There are no reports of acute or chronic effects in humans from exposure to *tert*-butyl hydroperoxide. Experimental studies of this chemical indicate that it will cause severe injury to the eyes and when placed in contact with the skin. It is moderately toxic by inhalation and ingestion and probably can cause irritation of the respiratory tract when inhaled.

The oral LD_{50} in rats is 460 mg/kg, and the LC_{50} (4-hour inhalation) in rats is 500 ppm. The percutaneous LD_{50} in the rat is 790 mg/kg. Five hundred mg of the chemical in contact with rabbit skin for 24 hours produced severe irritation. A 75% solution in dimethyl phthalate and a 35% solution of *tert*-butyl hydroperoxide in propylene glycol caused severe, permanent damage when applied to rabbit eyes. A 7% solution in propylene glycol caused slight irritation.

Special Handling Conditions

Tert-butyl hydroperoxide is very dangerous when exposed to heat or flame or by spontaneous chemical reaction. Slow first-order decomposition can be accelerated by the presence of 1 mole percent of copper, cobalt, or manganese salts. The substance can also react with reducing agents. There are currently no TLV or OSHA PEL related to this compound.

TRICHLOROETHYLENE

CAS Registry No.: 79-01-6

Synonyms: Triclene; ethene, trichloro

Structure

$$Cl_2C{=}C(Cl)H; \quad C_2HCl_3$$

Physical Properties

Molecular Weight: 131.4
Physical Form: Colorless liquid
Melting Point: −84.8°C
Boiling Point: 86.7°C
Flash Point: None
Specific Gravity (liquid): 1.456 (25°C/4°C)
Vapor Pressure: 19.9 mm Hg (0°C); 57.8 mm Hg (20°C)
Solubility: 0.105 g/100 ml water (20°C); soluble in ethanol and ethyl ether
Odor: Sweet; threshold = 21.4 ppm

Toxicity and Hazard

In humans, acute exposure to trichloroethylene primarily affects the central nervous system. Common symptoms are headache, dizziness, nausea, fatigue, and drunkenness. Coma and sudden death have been reported in severe intoxication. Exposure for 2 hours to 1000 ppm caused adverse affects to performance of steadiness and manual dexterity tests.

Acute skin exposure has caused erythema, burning, and inflammation of the skin. With repeated exposure, the liquid can produce inflammation and skin vesicles; in the eye, it produces pain and inflammation.

Trichloroethylene has a very low toxicity to animals via the inhalation and the oral routes: LC_{50} (mice, 4-hour inhalation) = 8450 ppm; LD_{50} (rats, oral) = 7.2 g/kg. Skin contact causes only mild irritation unless the affected area is closely occluded; skin absorption is not significant. When instilled into rabbit eyes, trichloroethylene has caused mild to moderate injury.

No significant changes were seen in rats exposed to 2000 ppm trichloroethylene, 5 days per week for 6 months. Female mice, however, exhibited fatty degeneration of the liver after exposure to 1600 ppm for up to 8 weeks. After inhaling 500-750 ppm for 3-8 weeks, dogs exhibited lethargy, anorexia, nausea, vomiting, weight loss, and liver dysfunction.

In a recent National Cancer Institute study, extremely high doses of trichloroethylene were shown to cause liver cancer in mice. There is no evidence that the material causes cancer in humans.

Special Handling Conditions

The TLV for trichloroethylene is 100 ppm (535 mg/m^3) [with a notice (1980) of intended change to 50 ppm] as an 8-hour time-weighted average; the exposure limit is 150 ppm over any 15-min period. The OSHA PEL for trichloroethylene is 100 ppm, with an acceptable ceiling of 200 ppm, and an acceptable maximum peak of 300 ppm for a duration of 5 min in any 2 hours. NIOSH (1973) has recommended an 8-hour time-weighted average of 100 ppm, with a ceiling of 150 ppm in any 10-min period.

Although trichloroethylene has caused cancer in mice, its carcinogenic potency is so low that no special precautions are needed for laboratory work with it beyond normal good practices. This includes the use of a hood for most operations.

VINYL CHLORIDE

CAS Registry No.: 75-01-4

Synonyms: Chloroethylene, chlorethene, VCM

Structure

$CH_2{=}CHCl$; C_2H_3Cl

Physical Properties

Molecular Weight: 62.50
Physical Form: Colorless gas at standard conditions but is usually handled as a liquid (sweet smelling)
Melting Point: −153.7°C
Boiling Point: −13.8°C
Flash Point: −78°C (C.O.C.)
Autoignition Temperature: 472°C
Flammable Limits: 3.6-33% by volume in air
Vapor Density (air = 1): 2.55 (25°C)
Liquid Density: 0.9013 g/ml (25°C)
Vapor Pressure: 2943 mm Hg (25°C)

Toxicity and Hazard

Brief exposure to vinyl chloride can produce dizziness and disorientation. With continued exposure, central nervous system depression can result, but exposure must be high (5000 ppm or more) to produce this effect. Because the gas is an irritant, there may be some coughing, but this effect is not prominent.

Chronic exposure to vinyl chloride at concentrations of 100 ppm or more has produced Raynaud's syndrome, lysis of the distal bones of the fingers, and a fibrosing dermatitis. These effects are probably related to continuous intimate contact with the skin by the substance.

Chronic exposure has also resulted in the production of a liver cancer—angiosarcoma—in about 50 workers worldwide (a relatively rare occurrence). All of these cancers have been produced by relatively high exposures, probably in the hundreds of parts per million range.

Vinyl chloride is flammable when exposed to heat, flame, or oxidizing agents. Large fires of the compound are very difficult to extinguish. Vapors represent a severe explosion hazard. Peroxides can be formed in air on standing and can explode. Phosgene is evolved on decomposition with heat.

Special Handling Conditions

Vinyl chloride is regulated by OSHA (29 CFR 1910.1017) as a human carcinogen. The regulations, which call for more stringent precautions than Procedure A (see Section I.B.8), should be consulted before work with vinyl chloride is begun. The allowable exposure limits are 1 ppm as an 8-hour time-weighted average and 5 ppm as a 15-min ceiling. Contact with liquid vinyl chloride is prohibited. A monitoring program is required for all vinyl chloride operations. Exposures to vinyl chloride must be controlled by feasible engineering controls or work practices. Use of closed systems or laboratory hoods that have protection factors adequate to prevent significant worker exposure are recommended. Whenever respirators are required, they must be used in accordance with a standard respirator program.

I.E.3 SELECTED BIBLIOGRAPHY—POTENTIAL HAZARDS OF KNOWN CHEMICALS

1. Altman, P. L.; Dittmer, D. S., Comp. and Ed., *Biology Data Book*, 2nd ed.; Fed. of Am. Societies for Exp. Biology: Bethesda, Md., 1972; Vols. I-III: Evaluation of reference data in the life sciences. Contains more than 18,000 literature citations.

2. Anon. *Toxic and Hazardous Industrial Chemicals Safety Manual*; International Technical Information Institute: Tokyo, 1979: Well-organized reference to specific materials, which highlights synonyms, properties, hazardous potentials (flammability, toxicity), handling and storage emergency measures, spill and leak procedures, and waste disposal; does not document its sources.

3. Bretherick, L. *Handbook of Reactive Chemical Hazards*, 2nd ed.; Butterworths: London-Boston, 1979: Contains data on stability, exposibility, flammability, and violent interactions between compounds for many laboratory chemicals.

4. Casarett, L. J.; Doull, J., Eds. *Toxicology*; Macmillan: New York, 1975: A complete and readable overview of toxicity, which includes metabolic pathways and relationships to related chemicals; good textbook but has not been arranged for ready reference for laboratory handling or emergencies.

5. Deichmann, W. B.; Gerarde, H. W. *Toxicology of Drugs and Chemicals*, 4th ed.; Academic Press: New York, 1969: A ready reference that presents data on side effects of drugs and toxicity of industrial chemicals.

6 *Fire Protection Guide on Hazardous Materials*, 7th ed.; National Fire Protection Assocation, Boston, Mass.: Contains NFPA-49-1975, hazard data on about 416 chemicals.

7. Goodman, L. S.; Gilman, A. *The Pharmacological Basis of Therapeutics*; Macmillan: New York, 1975: Contains toxicity information on drugs that affect many systems, water, salts and ions, gases and vapors, heavy metals, vitamins, and hormones.

8. Gosselin, R. E. *et al. Clinical Toxicology of Commercial Products: Acute Poisoning*, 4th ed.; Williams and Wilkins: Baltimore, 1976: "The purpose of this book is to assist the physician in dealing quickly and effectively with acute chemical poisonings, arising from misuse of commercial products." Contains trade names of products and their ingredients.

9. Hilado, C. J.; Clark, S. W. *Autoignition Temperatures of Organic Solvents*, Chem. Eng. (NY), 1972, 79(19), 75-80.

10. Lewis, R. J., Ed. *Registry of Toxic Effects of Chemical Substances*, DHEW (NIOSH), Publ. Microfiche issued quarterly: Contains data on more than 25,000

different chemicals. Useful as a first pass, but most of the data have not been evaluated by peer review.

11. Loomis, T. A. *Essentials of Toxicology*, 3rd ed.; Lea and Febiger: Philadelphia, 1978: A good basic primer for those interested in learning basic toxicology.

12. *NIOSH OSHA Product Guide to Chemical Hazards*; DHEW (NIOSH): Sept. 1978, Publ. No. 78-210: Presents, in tabular form, health hazards and protection measures for 380 specific chemicals for which there are federal regulations.

13. *Patty, F. A., Ed. Industrial Hygiene and Toxicology: Volume II—Toxicology*, Interscience Wiley: New York, 1963 (3rd ed. available 1980): A classical reference book that describes the toxicity of many different classes of organic and inorganic compounds.

14. Proctor, N.; Hughes, J. *Chemical Hazards in the Workplace*; Lippincott: Philadelphia, 1978: Excellent summary of toxic properties of laboratory materials; includes aids for diagnosis, treatment, and control; documents sources of information.

15. Sax, N. I. *Dangerous Properties of Industrial Materials*, 5th ed.; Van Nostrand-Reinhold: New York, 1979: Contains much data on fire and explosibility hazards, chemical reactivity, and toxicity of many chemicals.

16. M. Sittig. *Hazardous and Toxic Effects of Industrial Chemicals*. Noyes Data Corp.: Park Ridge, N.J., 1979: Excellent reference written with the industrial hygienist in mind; organized by material, giving physical state, synonyms, potential for exposure, exposure limits, route of entry into the body, harmful effects, medical surveillance, and protective methods; documents sources of information.

17. TLV Airborne Contaminants Committee. *TLVs®: Threshold Limit Values for Chemical Substances and Physical Agents in the Workroom Environment With Intended Changes*; American Conference of Governmental Industrial Hygienists: Cincinnati, Ohio, issued annually, and *Documentation of the Threshold Limit Values for Substances in Workroom Air and Supplemental Documentation*, American Conference of Governmental Industrial Hygienists: Cincinnati, Ohio (revised ed. expected 1980): Two excellent sources for the consideration of safe working levels.

18. Walters, D. B., Ed. *Safe Handling of Chemical Carcinogens, Mutagens, Teratogens, and Highly Toxic Substances*; Ann Arbor Science Publishers, Inc.: Ann Arbor, Mich.; 1980; Vol. 1.

19. Windholz, M., Ed. *The Merck Index*, 9th ed.; Merck and Company, Rahway, N.J., 1976: A classical reference book containing toxicity data on more than 9000 compounds.

20. Zabetakis, M. G. *Flammable Characteristics of Combustible Gases and Vapors*; U.S. Bureau of Mines Bulletin 627, 1965.

(See also Sections III.A.18 and III.E.7.)

I.F
Protective Apparel, Safety Equipment, Emergency Procedures, and First Aid

A variety of specialized clothing and equipment is commercially available for use in the laboratory. The proper use of these items will minimize or eliminate exposure to the hazards associated with many laboratory operations. The primary goal of laboratory safety procedures is the prevention of accidents and emergencies. However, accidents and emergencies may nonetheless occur and, at such times, proper safety equipment and correct emergency procedures can help minimize injuries or damage.

Every laboratory worker should be familiar with the location and proper use of the available protective apparel and safety equipment and with emergency procedures. Instruction on the proper use of such equipment, emergency procedures, and first aid should be available to everyone who might need it.

I.F.1 GLASSES AND FACE SHIELDS

General Eye-Protection Policy

Eye protection must be required for all personnel and any visitors present in locations where chemicals are stored or handled. No one should enter any laboratory without appropriate eye protection.

Conference rooms, libraries, offices, microscope rooms in which chemicals are not in use, and similar rooms are not normally eye-protection areas. However, at any time when chemicals are used in such rooms, even temporarily, signs should be posted and all persons in the vicinity warned that eye protection is temporarily required. For laboratory operations that do not involve the use of chemicals and, if chemicals are not used in the immediate vicinity, it may be permissible by arrangement with the laboratory supervisor, to remove the eye protection.

Safety spectacles that meet the criteria described below provide minimum eye protection for regular use. Additional protection may be required when carrying out more hazardous operations.

Laboratory management should make appropriate eye-protection devices available to visitors or others who only occasionally enter eye-protection areas. These devices would be used only on a temporary basis while the person is in the eye-protection area. (For example, it may be useful to place a container of inexpensive safety glasses next to the entrance to each laboratory for use by visitors.)

Contact lenses should not be worn in a laboratory. Gases and vapors can be concentrated under such lenses and cause permanent eye damage. Furthermore, in the event of a chemical splash into an eye, it is often nearly impossible to remove the contact lens to irrigate the eye because of involuntary spasm of the eyelid. Persons attempting to irrigate the eyes of an unconscious victim may not be aware of the presence of contact lenses, thus reducing the effectiveness of such treatment. Soft lenses can absorb solvent vapors even through face shields and, as a result, adhere to the eye.

There are some exceptional situations in which contact lenses must be worn for therapeutic reasons. Persons who must wear contact lenses should inform the laboratory supervisor so that satisfactory safety precautions can be devised.

Safety Spectacles

Ordinary prescription glasses do not provide adequate protection from injury to the eyes. The minimum acceptable eye protection requires the use of hardened-glass or plastic safety spectacles.

Safety glasses used in the laboratory should comply with the Standard for Occupational and Educational Eye and Face Protection (Z87.1) established by the American National Standards Institute. This standard specifies a minimum lens thickness of 3 mm, impact resistance requirements, passage of a flammability test, and lens-retaining frames.

Side shields that attach to regular safety spectacles offer some protection from objects that approach from the side but do not provide adequate protection from splashes. Other eye protection should be worn when a significant splash hazard exists.

Other Eye Protection

It is very important that each operation be analyzed to ensure that adequate eye protection is used. When operations that involve potential hazard to the eyes are performed (such as handling unusually corrosive chemicals), more complete eye protection than spectacles should be worn. It is the responsibility of the laboratory supervisor to determine the level of eye protection required and to enforce eye-protection rules.

Other forms of eye protection that may be required for a particular operation include the following:

Goggles

Goggles are not intended for general use. They are intended for wear when there is danger of splashing chemicals or flying particles. For example, goggles should be worn when working with glassware under reduced or elevated pressure and when glass apparatus is used in combustion or other high-temperature operations. Impact-protection goggles have screened areas on the sides to provide ventilation and reduce fogging of the lens and do not offer full protection against chemical splashes. Splash goggles ('acid goggles') (or face shields) that have splash-proof sides should be used when protection from harmful chemical splash is needed.

Face Shields

Goggles offer little protection to the face and neck. Full-face shields that protect the face and throat should always be worn when maximum protection from flying particles and harmful liquids is needed; for full protection, safety glasses should be worn with face shields. The metal-framed "nitrometer" mask offers greater protection for the head and throat from hazards such as flying glass or other light fragments. A face shield or mask may be needed when a vacuum system (which may implode) is used or when a reaction that has a potential for mild explosions is conducted.

Specialized Eye Protection

There are specific goggles and masks for protection against laser hazards and ultraviolet or other intense light sources, as well as glassblowing goggles and welding masks and goggles. The laboratory supervisor should determine whether the task being performed requires specialized eye protection and insist on the use of such equipment if it is necessary.

I.F.2 GLOVES

Skin contact is a potential source of exposure to toxic materials (see Section I.B.1); it is important that the proper steps be taken to prevent such contact.

1. Proper protective gloves (and other protective clothing, when necessary) should be worn whenever the potential for contact with corrosive or toxic materials and materials of unknown toxicity exists.
2. Gloves should be selected on the basis of the material being handled, the particular hazard involved, and their suitability for the operation being conducted.
3. Before each use, gloves should be inspected for discoloration, punctures, and tears.
4. Before removal, gloves should be washed appropriately. (NOTE: Some gloves, e.g., leather and polyvinyl alcohol, are water permeable.)
5. Glove materials are eventually permeated by chemicals. However, they can be used safely for limited time periods if specific use and glove characteristics (i.e., thickness and permeation rate and time) are known. Some of this information can be obtained from glove manufacturers, or the gloves used can be tested for breakthrough rates and times.
6. Gloves should be replaced periodically, depending on frequency of use and permeability to the substance(s) handled. Gloves overtly contaminated (if impermeable to water) should be rinsed and then carefully removed.

Gloves should be worn whenever it is necessary to handle corrosive materials, rough or sharp-edged objects, very hot or very cold materials, or whenever protection is needed against accidental exposure to chemicals. Gloves should not be worn around moving machinery. Many different types of gloves are commercially available.

1. Leather gloves may be used for handling broken glassware, for inserting glass tubes into rubber stoppers, and for similar operations where protection from chemicals is not needed.
2. There are various compositions and thicknesses of rubber gloves. Common glove materials include neoprene, polyvinyl chloride, nitrile, and butyl and natural rubbers. These materials differ in their resistance to various substances. [Specific information on this topic is often available from glove manufacturers' catalogs (although such data are usually only qualitative); example information is given in Table 3]. Rubber gloves should be inspected before each use; periodically, an inflation test, in

which the glove is first inflated with air and then immersed in water and examined for the presence of air bubbles, should be conducted.

3. Insulated gloves should be used when working at temperature extremes. Various synthetic materials such as Nomex® and Kevlar® can be used briefly up to 1000°F. Gloves made with these materials or in combination with other materials such as leather are available. It is best not to use gloves made either entirely or partly of asbestos, which is regulated as a carcinogen under OSHA, although such gloves probably do not present a great hazard.

4. Specialized gloves are manufactured for electrical linesmen, welders, and others. It is the responsibility of the laboratory supervisor to determine whether specialized hand protection is needed for any operation and to ensure that needed protection is available.

I.F.3 OTHER CLOTHING AND FOOTWEAR

The clothing worn by laboratory workers can be important to their safety. Such personnel should not wear loose (e.g., saris, dangling neckties, and overlarge or ragged laboratory coats), skimpy (e.g., shorts and/or halter tops), or torn clothing and unrestrained long hair. Loose or torn clothing and unrestrained long hair can easily catch fire, dip into chemicals, or become ensnarled in apparatus and moving machinery; skimpy clothing offers little protection to the skin in the event of chemical splash. If the possibility of chemical contamination exists, personal clothing that will be worn home should be covered by protective apparel.

Finger rings can react with chemicals and also should be avoided around equipment that has moving parts.

Shoes should be worn at all times in buildings where chemicals are stored or used. Perforated shoes, sandals, or cloth sneakers should not be worn in laboratories or areas where mechanical work is being done.

Protective Apparel

Appropriate protective apparel is advisable for most laboratory work and may be required for some. Such apparel can include laboratory coats and aprons, jump suits, special types of boots, shoe covers, and gauntlets. It can be either washable or disposable in nature. Garments are commercially available that can help protect the laboratory worker against chemical splashes or spills, heat, cold, moisture, and radiation.

Protective apparel should resist physical hazards and permit easy execution of manual tasks while being worn. It should also satisfy other performance requirements—strength, chemical and thermal resistance,

TABLE 3 Resistance to Chemicals of Common Glove Materials (E = Excellent, G = Good, F = Fair, P = Poor)

Chemical	Natural Rubber	Neoprene	Nitrile	Vinyl
Acetaldehyde	G	G	E	G
Acetic acid	E	E	E	E
Acetone	G	G	G	F
Acrylonitrile	P	G	–	F
Ammonium hydroxide (sat)	G	E	E	E
Aniline	F	G	E	G
Benzaldehyde	F	F	E	G
Benzene[a]	P	F	G	F
Benzyl chloride[a]	F	P	G	P
Bromine	G	G	–	G
Butane	P	E	–	P
Butyraldehyde	P	G	–	G
Calcium hypochlorite	P	G	G	G
Carbon disulfide	P	P	G	F
Carbon tetrachloride[a]	P	F	G	F
Chlorine	G	G	–	G
Chloroacetone	F	E	–	P
Chloroform[a]	P	F	G	P
Chromic acid	P	F	F	E
Cyclohexane	F	E	–	P
Dibenzyl ether	F	G	–	P
Dibutyl phthalate	F	G	–	P
Diethanolamine	F	E	–	E
Diethyl ether	F	G	E	P
Dimethyl sulfoxide[b]	–	–	–	–
Ethyl acetate	F	G	G	F
Ethylene dichloride[a]	P	F	G	P
Ethylene glycol	G	G	E	E
Ethylene trichloride[a]	P	P	–	P
Fluorine	G	G	–	G
Formaldehyde	G	E	E	E
Formic acid	G	E	E	E
Glycerol	G	G	E	E
Hexane	P	E	–	P
Hydrobromic acid (40%)	G	E	–	E

TABLE 3 (*continued*)

Chemical	Natural Rubber	Neoprene	Nitrile	Vinyl
Hydrochloric acid (conc)	G	G	G	E
Hydrofluoric acid (30%)	G	G	G	E
Hydrogen peroxide	G	G	G	E
Iodine	G	G	–	G
Methylamine	G	G	E	E
Methyl cellosolve	F	E	–	P
Methyl chloride[a]	P	E	–	P
Methyl ethyl ketone	F	G	G	P
Methylene chloride[a]	F	F	G	F
Monoethanolamine	F	E	–	E
Morpholine	F	E	–	E
Naphthalene[a]	G	G	E	G
Nitric acid (conc)	P	P	P	G
Perchloric acid	F	G	F	E
Phenol	G	E	–	E
Phosphoric acid	G	E	–	E
Potassium hydroxide (sat)	G	G	G	E
Propylene dichloride[a]	P	F	–	P
Sodium hydroxide	G	G	G	E
Sodium hypochlorite	G	P	F	G
Sulfuric acid (conc)	G	G	F	G
Toluene[a]	P	F	G	F
Trichloroethylene[a]	P	F	G	F
Tricresyl phosphate	P	F	–	F
Triethanolamine	F	E	E	E
Trinitrotoluene	P	E	–	P

[a]Aromatic and halogenated hydrocarbons will attack all types of natural and synthetic glove materials. Should swelling occur, the user should change to fresh gloves and allow the swollen gloves to dry and return to normal.

[b]No data on the resistance to dimethyl sulfoxide of natural rubber, neoprene, nitrile rubber, or vinyl materials are available; the manufacturer of the substance recommends the use of butyl rubber gloves.

flexibility, and ease of cleaning. The required degree of performance can be determined on the basis of the substances being handled. The choice of garment—laboratory coat versus rubber or plastic apron versus disposable jump suit—depends on the degree of protection required and is the responsibility of the supervisor.

Laboratory coats are intended to prevent contact with dirt and the minor chemical splashes or spills encountered in laboratory-scale work. The cloth laboratory coat is, however, primarily a protection for clothing and may itself present a hazard (e.g., combustibility) to the wearer; cotton and synthetic materials such as Nomex or Tyvek® are satisfactory; rayon and polyesters are not. Laboratory coats do not significantly resist penetration by organic liquids and, if significantly contaminated by them, should be removed immediately.

Plastic or *rubber aprons* provide better protection from corrosive or irritating liquids but can complicate injuries in the event of fire. Furthermore, a plastic apron can accumulate a considerable charge of static electricity and should be avoided in areas where flammable solvents or other materials could be ignited by a static discharge.

Disposable outer garments (e.g., Tyvek) may, in some cases, be preferable to reusable ones. One such case is that of handling appreciable quantities of known carcinogenic materials (see Section I.B.8), for which long sleeves and the use of gloves are also recommended. Disposable full-length jump suits are strongly recommended for high-risk situations, which may also require the use of head and shoe covers. Many disposable garments, however, offer only limited protection from vapor penetration and considerable judgment is needed when using them. Impervious suits fully enclosing the body may be necessary in emergency situations.

Laboratory workers should know the appropriate techniques for removing protective apparel, especially any that has become contaminated. Chemical spills on leather clothing or accessories (watchbands, shoes, belts, and such) can be especially hazardous because many chemicals can be absorbed in the leather and then held close to the skin for long periods. Such items must be removed promptly and decontaminated or discarded to prevent the possibility of chemical burn (see Chapter II.E). Specialized or disposable clothing for use with particular classes of hazardous chemicals should be treated in a similar way (see Chapters II.E, F, and G).

Safety showers (see Section I.F.4) should be readily accessible for use when a chemical spill contaminates large sections of clothing or skin.

Foot Protection

More extensive foot protection than ordinary shoes may be required in some cases. Rubber boots or plastic shoe covers may be used to avoid possible exposure of the feet to corrosive chemicals or large quantities of solvents and water that might penetrate normal foot gear (e.g., during cleanup operations). Because these types of boots and covers may increase the risk of static spark, their use in normal laboratory operations is not advisable.

Other specialized tasks may require footwear that has, for example, conductive soles, insulated soles, or built-in metal toe caps. The laboratory supervisor should recommend the use of such protection whenever appropriate.

I.F.4 SAFETY EQUIPMENT

Safety and emergency equipment should be available in all laboratories. The protection afforded by this equipment depends on its proper and consistent use. Laboratory workers should realize that safety devices are intended to help protect them from injury and should not avoid using such devices when they are needed.

All laboratories in which chemicals are used should have available fire extinguishers, safety showers, and eyewash fountains, as well as laboratory hoods and laboratory sinks (which can be considered part of the safety equipment of the laboratory); (see Section I.H.2 for a discussion of laboratory hoods). Respiratory protection for emergency use should be available nearby, along with fire alarms, emergency telephones, and identified emergency telephone numbers.

In addition to these standard items, there may also be a need for other protection. It is the responsibility of the laboratory supervisor to recommend and provide supplementary safety equipment as needed. The special precautions needed when dealing with DNA biological experiments and with radioisotopes are not within the scope of this report; such material can be found elsewhere [*Guidelines for Research Involving Recombinant DNA Molecules* 45 Fed. Reg. 25,366-25,370 (1980) and U.S. Nuclear Regulatory Commission *Rules and Regulations*, 10 CFR Chap. 1].

FIRE SAFETY EQUIPMENT

Typical fire safety equipment in the laboratory includes a variety of fire extinguishers and may include fire hoses, blankets, and automatic extinguishing systems.

Fire Extinguishers

All chemical laboratories should be provided with carbon dioxide or dry chemical fire extinguishers (or both). Other types of extinguishers should be available if required by the work being done. The four types of extinguishers most commonly used are classified by the type of fire (see below) for which they are suitable.

1. Water extinguishers are effective against burning paper and trash (Class A fires). These should not be used for extinguishing electrical, liquid, or metal fires.

2. Carbon dioxide extinguishers are effective against burning liquids, such as hydrocarbons or paint, and electrical fires (Classes B and C fires). They are recommended for fires involving delicate instruments and optical systems because they do not damage such equipment. They are less effective against paper and trash or metal fires and should not be used against lithium aluminum hydride fires.

3. Dry powder extinguishers, which contain sodium bicarbonate, are effective against burning liquids and electrical fires (Classes B and C fires). They are less effective against paper and trash or metal fires. They are not recommended for fires involving delicate instruments or optical systems because of the cleanup problem this creates. These extinguishers are generally used where large quantities of solvent may be present.

4. Met-L-X® extinguishers and others that have special granular formulations are effective against burning metal (Class D fires). Included in this category are fires involving magnesium, lithium, sodium, and potassium; alloys of reactive metals; and metal hydrides, metal alkyls, and other organometallics. These extinguishers are less effective against paper and trash, liquid, or electrical fires.

Every extinguisher should carry a label indicating what class or classes of fires it is effective against. There are a number of other more specialized types of extinguishers available for unusual fire hazard situations. Each laboratory worker should be responsible for knowing the location, operation, and limitations of the fire extinguishers in the work area. It is the responsibility of the laboratory supervisor to ensure that all laboratory workers are shown the locations of nearby fire extinguishers and are trained in their use. After use, an extinguisher should be recharged or replaced by designated personnel.

Fire Hoses

Fire hoses are intended for use by trained firefighting personnel against fires too large to be handled by extinguishers but are included as safety equipment in some structures. Water has a cooling action and is effective against fires involving paper, wood, rags, trash, and such (Class A fires). Water should not be used directly on fires that involve live electrical equipment (Class C fires) or chemicals such as alkali metals, metal hydrides, and metal alkyls that react vigorously with it (Class D fires).

Streams of water should not be used against fires that involve oils or other water-insoluble flammable liquids (Class B fires). This form of water will not readily extinguish such fires, and it will usually spread or float the fire to adjacent areas. These possibilities are minimized by the use of a water fog.

Water fogs are used extensively by the petroleum industry because of their fire-controlling and extinguishing properties. A fog can be used safely and effectively against fires that involve oil products, as well as those involving wood, rags, rubbish, and such.

Because of the potential hazards in using water around chemicals (and the problem of controlling a hose delivering water at significant pressures), laboratory workers should refrain from using fire hoses except in extreme emergencies. Such use should be reserved for trained firefighting personnel.

Fire Blankets

Many laboratories still have fire blankets available. A fire blanket is used primarily as a first aid measure for the prevention of shock rather than against smoldering or burning clothing. It should be used only as a last-resort measure to extinguish clothing fires: such blankets tend to hold heat in and increase the severity of burns. Clothing fires should be extinguished by immediately dropping to the floor and rolling or, if a safety shower (see below) is immediately available, using it.

Automatic Fire-Extinguishing Systems

In areas where fire potential (for example, solvent storage areas) and the risk of injury or damage are high, automatic fire-extinguishing systems are often used. These may be of the water-sprinkler, carbon dioxide, dry chemical, or halogenated hydrocarbon types. Whenever it has been determined that the risk justifies an automatic fire-extinguishing system, the laboratory workers should be informed of its presence and advised of any safety precautions required for its action (e.g., evacuation before a carbon dioxide total-flood system is actuated).

Respiratory Protective Equipment

The primary method for the protection of laboratory personnel from airborne contaminants should be to minimize the amount of such materials entering the laboratory air (see Chapter I.B.8). When effective engineering controls are not possible, suitable respiratory protection should be provided. It is the responsibility of the laboratory supervisor to determine when such protection is needed and to ensure that it is used.

Under OSHA regulations, only equipment listed and approved by the Mine Safety and Health Administration (MSHA) and the National Institute for Occupational Safety and Health (NIOSH) may be used for respiratory protection. Also under the regulations, each site on which respiratory protective equipment is used must implement a respirator program in compliance with the standard (29 CFR 1910.134) (see also ANSI 288.2).

Types of Respirators

Several types of nonemergency respirators are available for protection in atmospheres that are not immediately dangerous to life or health but could be detrimental after prolonged or repeated exposure. Other types of respirators are available for emergency or rescue work in atmospheres from which the wearer cannot escape without respiratory protection. In either case, additional protection may be required if the airborne contaminant is of a type that could be absorbed through or irritate the skin. For example, the possibility of eye or skin irritation may require the use of a full-body suit and a full-face mask rather than a half-face mask.

The choice of the appropriate respirator to use in a given situation will depend on the type of contaminant and its estimated or measured concentration, known exposure limits, and warning and hazardous properties (e.g., eye irritation or skin absorption). Once this information is available, an appropriate type of respirator can be selected by using a guide such as Table 4. The degree of protection afforded by the respirator varies with the type.

1. *Chemical cartridge respirators*—These can be used only for protection against particular individual (or classes of) vapors or gases as specified by the respirator manufacturer and cannot be used at concentrations of contaminant above that specified on the cartridge. Also, these respirators cannot be used if the oxygen content of the air is less than 19.5%, in atmospheres immediately dangerous to life, or for rescue or emergency work. These respirators function by the entrapment of vapors and gases in a cartridge or canister that contains a sorbent material. Activated charcoal is probably the most common adsorbent. Because it is possible for significant breakthrough to occur at a fraction of the canister capacity, knowledge of the potential workplace exposure and length of time the respirator will be worn is important. It may be desirable to replace the cartridge after each use to ensure the maximum available exposure time

TABLE 4 Guide for Selection of Respirators

Type of Hazard	Type of Respirator
Oxygen deficiency	Self-contained breathing apparatus Hose mask with blower Combination of air-line respirator and auxiliary self-contained air supply or air-storage receiver with alarm
Gas and vapor contaminants Immediately dangerous to life or health	Self-contained breathing apparatus Hose mask with blower Air-purifying full-facepiece respirator with chemical canister (gas mask) Self-rescue mouthpiece respirator (for escape only) Combination of air-line respirator and auxiliary self-contained air supply or air-storage receiver with alarm
Not immediately dangerous to life or health	Air-line respirator Hose mask with blower Air-purifying half-mask or mouthpiece respirator with chemical cartridge
Particulate Contaminants Immediately dangerous to life or health	Self-contained breathing apparatus Hose mask with blower Air-purifying full-facepiece respirator with appropriate filter Self-rescue mouthpiece respirator (for escape only) Combination of air-line respirator and auxiliary self-contained air supply or air-storage receiver with alarm
Not immediately dangerous to life or health	Air-purifying half-mask or mouthpiece respirator with filter pad or cartridge Air-line respirator Air-line abrasive-blasting respirator Hose mask with blower
Combination of gas, vapor, and particulate contaminants Immediately dangerous to life or health	Self-contained breathing apparatus Hose mask with blower Air-purifying full-facepiece respirator with chemical canister and appropriate filter (gas mask with filter) Self-rescue mouthpiece respirator (for escape only) Combination of air-line respirator and auxiliary self-contained air supply or air-storage receiver with alarm
Not immediately dangerous to life or health	Air-line respirator Hose mask without blower Air-purifying half-mask or mouthpiece respirator with chemical cartridge and appropriate filter

Source: ANSI Standard Z88.2 (1969).

for each new use. Difficulty in breathing or the detection of odors indicates plugged or exhausted filters or cartridges or concentrations of contaminants higher than the absorbing capability of the cartridge; such items should be replaced promptly (if necessary, by a more effective type of respirator).

These respirators must fit snugly to the face to be effective. Conditions that prevent facepiece-to-face seal (for example, temple pieces of glasses or facial hair) will permit contaminated air to bypass the filter, possibly creating a dangerous situation for the user. Tests for the proper fit of the respirator on the user should be conducted prior to its selection and verified before he or she enters the area of contamination.

Organic vapor cartridges cannot be used for vapors that have poor warning properties or those that will generate high heats of reaction with the sorbent materials in the cartridge.

2. *Dust, fume, and mist respirators*—These can be used only for protection against particular individual (or classes of) dusts, fumes, and mists as specified by the manufacturer. Such particulate-removing respirators usually trap the particles in a filter composed of fibers. Respirators of this type are generally disposable. Some examples are surgical masks and 3M® toxic-dust and nuisance-dust masks, which can be used to filter out animal dander and nontoxic and nuisance dusts. Some are NIOSH-approved for more specific purposes such as protection against simple or benign dust and fibrogenic dusts and asbestos. They are not 100% efficient in removing particles. The useful life of the filter is dependent on the concentration of contaminant encountered.

Particulate-removing respirators such as surgical masks afford no protection against gases or vapors and should not be used when handling chemicals. They provide little if any protection and may give the user a false sense of security. They are also subject to the limitations of fit described above.

3. *Supplied-air respirators*—These supply fresh air to the facepiece of the respirator at a pressure high enough to cause a slight buildup relative to atmospheric. As a result, the supplied air flows outward from the mask and contaminated air from the work environment cannot readily enter the mask. This characteristic renders face-to-facepiece fit less important than with other types of respirators. Fit testing is, however, required before selection and use.

Supplied-air respirators are effective protection against a wide range of air contaminants (gases, vapors, and particulates) and can be used where oxygen-deficient atmospheres are present. They are free from the maintenance problems associated with charcoal, particulate, and chemical-scrubbing filters. Such respirators can be used where concentrations of air contaminants could be immediately dangerous to life or from which the wearer could not escape unharmed without the air of the respirator provided (a) the protection factor of the respirator is not exceeded and (b) the provisions of 29 CFR 1910.134 (a safety harness and an escape system in case of compressor failure) are not violated.

The air supply of this type of respirator must be kept free of contaminants (e.g., by use of oil filters and CO absorbers), and some consideration should be given to its quality and relative humidity. Most laboratory air is not suitable for use with these units. These units usually require the user to drag long lengths of hose connected to the air supply. Thus, the range of their use is limited to the maximum length of hose specified by the manufacturer. Some supplied-air units are used in combination with self-contained apparatus.

4. *Self-contained breathing apparatus*—This is the only type of respiratory protective equipment suitable for emergency or rescue work. This equipment consists of a full-face mask connected to a cylinder of compressed air and has no limitations with regard to its use in areas of toxic contaminants or oxygen deficiency. However, the air supply is limited to the capacity of the cylinder (5- to 30-min use time) and, therefore, these respirators cannot be used for extended periods without recharging or replacing the cylinders; for safety reasons, the "pressure/demand" type, which always has a positive pressure within the mask, is much preferred to the "demand" type. Also, they are bulky and heavy, and additional protective apparel may still be required depending on the nature of the hazard. All institutions or organizations that have laboratories in which chemicals are used should have protective equipment of this type available for emergencies and provide training in its use to selected personnel.

Procedures and Training

Each area where respirators are used should have written information available that shows the limitations, fitting methods, and inspection and cleaning procedures for each type of respirator available. Personnel who may have occasion to use respirators in their work must be thoroughly trained in the fit testing, use, limitations, and care of such equipment. Training should include demonstrations and practice in wearing, adjusting, and properly fitting the equipment. Contact lenses should not be worn when a respirator is used, especially in a highly contaminated area. OSHA regulations require that a worker be examined by a physician before beginning work in an area where a respirator must be worn [29 CFR 1910.134(b)(10)].

Inspections

Respirators for routine use should be inspected before each use by the user and periodically by the laboratory supervisor. Self-contained breathing apparatus should be inspected once a month and cleaned after each use. Defective units should not be used but should be repaired by a qualified person or replaced promptly.

Safety Showers and Eyewash Fountains

Safety Showers

Safety showers should be provided in areas where chemicals are handled for immediate first aid treatment of chemical splashes and for extinguishing clothing fires. Every laboratory worker should learn the locations of and how to use the safety showers in the work area so that he or she can find them with eyes closed, if necessary. Safety showers should be tested routinely by laboratory personnel to ensure that the valve is operable and to remove any debris in the system.

The shower should be capable of drenching the subject immediately and should be large enough to accommodate more than one person if necessary. It should have a quick-opening valve requiring manual closing; a downward-pull delta bar is satisfactory if long enough but chain pulls are not advisable because of the potential for persons to be hit by them and the difficulty of grasping them in an emergency.

Eyewash Fountains

Eyewash fountains may be required if the substance in use presents an eye hazard or in research or instructional laboratories where unknown hazards may be encountered. An eyewash fountain should provide a soft stream or spray of aerated water for an extended period (15 min). These fountains should be located close to the safety showers so that, if necessary, the eyes can be washed while the body is showered.

Other Safety Equipment

Safety Shields

Safety shields should be used for protection against possible explosions or splash hazards. Laboratory equipment should be shielded on all sides so that there is no line-of-sight exposure of personnel.

Provided its opening is covered by closed doors, the conventional laboratory exhaust hood is a readily available built-in shield. However, a portable shield should also be used when manipulations are performed, particularly with hoods that have vertical-rising doors rather than horizontal-sliding sashes.

Portable shields can be used to protect against hazards of limited severity, e.g., small splashes, heat, and fires. A portable shield, however, provides no protection at the sides or back of the equipment and many such shields are not sufficiently weighted and may topple toward the worker when there is a blast (perhaps hitting him or her and also permitting exposure to flying objects). A fixed shield that completely surrounds the experimental apparatus can afford protection against minor blast damage.

Methyl methacrylate, polycarbonate, polyvinyl chloride, and laminated safety plate glass are all satisfactory transparent shielding materials. Where combustion is possible, the shielding material should be nonflammable or slow burning; if it can withstand the working blast pressure, laminated safety plate glass may be the best material for such circumstances. When cost, transparency, high tensile strength, resistance to bending loads, impact strength, shatter resistance, and burning rate are considered, methyl methacrylate offers an excellent overall combination of shielding characteristics. Polycarbonate is much stronger and self-extinguishing after ignition but is readily attacked by organic solvents (see also Section I.C.2).

Stretchers

Although stretchers are often provided in areas where chemicals are handled, untrained personnel should use them only in life-threatening situations. It is generally best not to move a seriously injured person until qualified medical help arrives.

Storage and Inspection of Emergency Equipment

It is often useful to establish a central location for storage of emergency equipment. Such a location should contain the following:

1. self-contained breathing apparatus,
2. safety belt with rope (to maintain contact with rescuers entering a laboratory under emergency conditions),
3. blankets for covering injured persons,
4. stretchers, and

5. first aid equipment (for unusual situations such as exposure to cyanide, where immediate first aid is required).

Safety equipment should be inspected regularly (e.g., every 3 to 6 months) to ensure that it will function properly when needed. It is the responsibility of the laboratory supervisor or safety coordinator to establish a routine inspection system and to verify that inspection records are kept.

1. Fire extinguishers should be inspected for broken seals, damage, and low gage pressure (depending on type of extinguisher). Proper mounting of the extinguisher and its ready accessibility should also be checked. Some types of extinguishers must be weighed annually, and periodic hydrostatic testing may be required.
2. Self-contained breathing apparatus should be checked at least once a month (and after each use) to determine whether proper air pressure is being maintained. The examiner should look for signs of deterioration or wear of rubber parts, harness, and hardware and make certain that the apparatus is clean and free of visible contamination.
3. Safety showers and eyewash fountains should be examined visually and their mechanical function should be tested.

I.F.5 EMERGENCY PROCEDURES

The following emergency procedures are recommended in the event of fire, explosion, or other laboratory accident. These procedures are intended to limit injuries and minimize damage if an accident should occur.

1. Render assistance to persons involved and remove them from exposure to further injury if necessary.
2. Warn personnel in adjacent areas of any potential hazards to their safety.
3. Render immediate first aid; appropriate measures include washing under a safety shower, administering oxygen and artificial resuscitation if breathing has stopped, and special first aid measures (such as the use of a cyanide first aid kit if cyanide exposure is involved).
4. Extinguish small fires by using a portable extinguisher. Turn off nearby apparatus and remove combustible materials from the area. In case of larger fires, contact the appropriate fire department promptly.

In case of medical emergency, laboratory personnel should remain calm and do only what is necessary to protect life.

1. Summon medical help immediately.

2. Do not move an injured person unless he or she is in danger of further harm.

3. Keep any injured person warm. If feasible, designate one person to remain with the injured person. The injured person should be within sight, sound, or physical contact of that person at all times.

4. If clothing is on fire, knock the person to the floor and roll him or her around to smother the flames or, if a safety shower is immediately available, douse the person with water.

5. If chemicals have been spilled on the body, flood the exposed area with sufficient running water from the safety shower and immediately remove any contaminated clothing.

6. If a chemical has entered the eye, immediately wash the eyeball and the inner surface of the eyelid with plenty of water for 15 min. An eyewash fountain should be used if available. Forcibly hold the eye open to wash thoroughly behind the eyelids.

Preparing for Emergencies

It is the responsibility of every laboratory organization to establish a specific emergency plan for its facilities. Such a plan should include evacuation routes and shelter areas, medical facilities, and procedures for reporting all accidents and emergencies and should be reinforced by frequent drills and simulated emergencies.

Evacuation Procedures

Evacuation procedures should be established and communicated to all personnel.

1. *Emergency alarm system*—A system should be available to alert personnel in the event of an emergency that may require evacuation. Laboratory personnel should be familiar with the location and operation of this equipment. A system should be established to relay telephone alert messages; the names and telephone numbers of personnel responsible for each laboratory or other area should be prominently posted in case of emergencies outside regular working hours.

Isolation areas (e.g., cold, warm, or sterile rooms) should be equipped with alarm or telephone systems that can be used to alert outsiders to the presence of a worker trapped inside or to warn workers inside of the existence of an emergency outside that requires their evacuation. Where unusually toxic substances are handled, it may be desirable to have a monitoring and alarm system so that, if the concentration of the substances in the work environment exceeds a set limit, an alarm is sounded to warn the laboratory workers to evacuate the area.

2. *Evacuation procedures*—Evacuation routes and alternatives may need to be established and, if so, should be communicated to all personnel. An outside assembly area for evacuated personnel should be designated.

3. *Shutdown procedures*—Brief guidelines for shutting down operations during an emergency or evacuation should be communicated to all personnel.

4. *Return and start-up procedures*—Return procedures to ensure that personnel do not return to the laboratory until the emergency is ended and start-up procedures that may be required for some operations should be prominently displayed and reviewed regularly.

5. *Drills*—All aspects of the emergency procedure should be tested regularly (e.g., every 6 months to a year), and trials of evacuations (if there are such procedures) should be held periodically.

Medical Facilities

Laboratories that do not have a regular medical staff should have personnel trained in first aid available during regular working hours to render assistance until medical help can be obtained.

An emergency room staffed with medical personnel specifically trained in proper treatment of chemical exposure should be readily accessible. For small laboratories, prior arrangement with a nearby hospital or emergency room may be necessary to ensure that treatment will be available promptly. The services of an ophthalmologist especially alerted to and familiar with chemical injury treatment should also be available to minimize the damage to eyes that may result from many types of laboratory incidents. Proper and speedy transportation of the injured to the medical treatment facility should be available. In addition to the normal facilities found in an emergency room, there should be special provisions that include specific standing orders for emergency treatment of chemical accidents. Emergencies that should be anticipated include the following:

1. thermal and chemical burns;
2. cuts and puncture wounds from glass or metal, including possible chemical contamination;
3. skin irritation by chemicals;

4. poisoning by ingestion, inhalation, or skin absorption;
5. asphyxiation (chemical or electrical); and
6. injuries to the eyes from splashed chemicals.

Accident and Emergency Reporting

A system should be established to ensure that accidents or emergencies are promptly reported to the persons responsible for safety matters. Such reports are required by law in many cases and help to uncover hazards that can be corrected. In all cases, the report should be in a written form and retained as a part of the safety record program.

Fires and Explosions

Small fires that can easily be extinguished without evacuating the building or calling the fire department are among the most common laboratory incidents. However, even a minor fire can quickly become a serious problem. The first few minutes after discovery of a fire are critical in preventing a larger emergency. The following actions should be taken by laboratory personnel in case of a minor fire:

1. Alert other personnel in the laboratory and send someone for assistance.

2. Attack the fire immediately, but never attempt to fight a fire alone. A fire in a small vessel can often be suffocated by covering the vessel with an inverted beaker or a watch glass. Use the proper extinguisher, directing the discharge of the extinguisher at the base of the flame. To ensure that the proper type of extinguisher is used (see Section I.F.4), compare its use label with the descriptions below.

Class A fires—ordinary combustible solids such as paper, wood, coal, rubber, and textiles;

Class B fires—petroleum hydrocarbons (diesel fuel, motor oil, and grease) and volatile flammable solvents;

Class C fires—electrical equipment;

Class D fires—combustible or reactive metals (such as sodium and potassium), metal hydrides, or organometallics (such as alkylaluminums).

3. Avoid entrapment in a fire; always fight a fire from a position accessible to an exit.

If there is any doubt whether the fire can be controlled by locally available personnel and equipment, the following actions should be taken:

1. Notify the fire department and activate the emergency alarm system.
2. Confine the emergency (close hood sashes, doors between laboratories, and fire doors) to prevent further spread of the fire.
3. Assist injured personnel (provide first aid or transportation to medical aid if necessary) (see Section I.F.6).
4. Evacuate the building to avoid further danger to personnel.

In case of an explosion, immediately turn off burners and other heating devices, stop any reactions in progress, assist in treating victims, and vacate the area until it has been decontaminated.

It is the responsibility of the laboratory supervisor to determine whether unusual hazards exist that require more stringent safety precautions. In large laboratories, or where risk is high, designated firefighting teams may be necessary to minimize risk. Special arrangements with local fire departments to warn them of the hazards of chemical fires may be desirable in some situations.

Chemical Spills

For most small-scale laboratory spills, the procedures described in Section II.E.6 will be adequate. Where large-scale spills may be possible, emergency procedures should be prepared for containing spilled chemicals with minimal damage. A spill-control policy should include consideration of the following points:

1. *prevention*—storage, operating procedures, monitoring, inspection, and personnel training;
2. *containment*—engineering controls on storage facilities and equipment;
3. *cleanup*—countermeasures and training of designated personnel to help reduce the impact of a chemical spill; and
4. *reporting*—provisions for reporting spills both internally (to identify controllable hazards) and externally (for example, to state and federal regulatory agencies).

Other Emergencies

Laboratories should be prepared for hazards resulting from loss of any utility service or severe weather. Loss of the water supply, for example, can render safety showers, eyewash fountains, and sprinkler systems inoperative. All hazardous laboratory work should cease until service is restored.

I.F.6 FIRST AID

First aid is the immediate care of a person who has been injured or has suddenly taken ill. It is intended to prevent death or further illness and injury and to relieve pain until medical aid can be obtained. The objectives of first aid are (1) to control conditions that might endanger life; (2) to prevent further injury; (3) to relieve pain, prevent contamination, and treat for shock; and (4) to make the patient as comfortable as possible.

The initial responsibility for first aid rests with the first person(s) at the scene, who should react quickly but in a calm and reassuring manner. The person assuming responsibility should immediately summon medical help (be explicit in reporting suspected types of injury or illness and requesting assistance). The injured person should not be moved except where necessary to prevent further injury. Laboratory workers should be encouraged to obtain training in first aid and cardiopulmonary resuscitation (CPR).

Pulmonary Resuscitation

If the patient is unresponsive and no breathing movements are apparent, begin mouth-to-mouth resuscitation immediately. Delay increases the risk of serious disability or death.

1. Place the patient flat on his or her back on the floor and kneel at the side.
2. Establish an airway. Check the patient's mouth with your finger to be sure that no obstruction is present and then tip the patient's head back until the chin points straight up.
3. Pinch the patient's nostrils, and begin mouth-to-mouth resuscitation by taking a deep breath and placing your mouth over the patient's mouth so as to make a leakproof seal. Blow your breath into the patient's mouth until you see the chest rise.
4. Remove your mouth and allow the patient to exhale.
5. Repeat the procedure at a rate of once every 5 s.

Heart (Cardiac) Resuscitation

In the unresponsive patient, check for a cardiac pulse; locate the larynx or adam's apple with the tips of the fingers and slide them into the groove between it and the muscle at the side of the neck. If no pulse is felt, circulation must be reestablished within 4 min to prevent brain damage.

1. With the patient flat on his or her back, kneel at the waist, facing the head.
2. Place the heel of your right hand over the heel of your left hand on top of the patient's breastbone about 3 cm above its lower tip.
3. Shift your weight to the patient's chest and compress it at least 4 cm, then remove the pressure.
4. Continue at a rate of 80 times/min.

Heavy Bleeding

Heavy bleeding is caused by injury to one or more large blood vessels. Lay the patient down. Control bleeding by applying firm pressure directly over the wound with a clean handkerchief, cloth, or your hand. A tourniquet should be applied only in cases of an amputation or other injury to a limb in which there is no other way to stop the bleeding. If a tourniquet is used, a record of the time it was applied must be kept.

Shock

Shock usually accompanies severe injury. The signs of shock include pallor, a cold and clammy skin and beads of perspiration on the forehead and palms or hands, weakness, nausea or vomiting, shallow breathing, and a rapid pulse that may be too faint to be felt at the wrist. The following procedures for the treatment of shock should be followed:

1. Correct the cause, if possible (e.g., control bleeding).
2. Keep the patient lying down; if there are no contraindications (e.g., a head injury), elevate the patient's legs.
3. Keep the patient's airway open. If he or she is about to vomit, turn the head to the side.
4. Keep the patient warm.

Miscellaneous Illnesses and Injuries

After requesting medical aid, the following points should be addressed in specific emergencies:

1. *Abdominal pain*—Keep the patient quiet. Give nothing by mouth.
2. *Back and neck injuries*—Keep the patient absolutely quiet. Do not move the patient or lift the head unless absolutely necessary.
3. *Chest pain*—Keep the patient calm and quiet. Place the patient in the most comfortable position (usually half sitting).

4. *Convulsion or epileptic seizure*—Place the patient on the floor or a couch. Do not restrain the patient's movements except to prevent injury. Do not place a blunt object between the teeth, put any liquid in the mouth, or slap the patient or douse him or her with water.

5. *Electric shock*—Throw the switch to turn off the current. Do not touch the victim until he or she is separated from the current source. Begin mouth-to-mouth resuscitation if respiration has ceased.

6. *Fainting*—Simple fainting can usually be ended quickly by laying the victim down.

7. *Unexplained unconsciousness*—Look for emergency medical identification around the victim's neck or wrist or in his or her wallet. Keep the victim warm, lying down, and quiet until he or she regains consciousness. Do not move the victim's head if there is bleeding from the nose, mouth, ear, or eyes. Do not give the victim anything by mouth. Keep the victim's airway open to aid breathing. Do not cramp the neck with a pillow.

I.F.7 CHEMICAL INGESTION OR CONTAMINATION

Ingestion of Chemicals

Encourage the victim to drink large amounts of water while en route to medical assistance. Attempt to learn exactly what substances were ingested and inform the medical staff (while the victim is en route, if possible) and the local poison control center.

Chemicals Spilled on the Body over a Large Area

Quickly remove all contaminated clothing while using the safety shower; seconds count and no time should be wasted because of modesty. Immediately flood the affected body area with cold water for at least 15 min; resume if pain returns. Wash off chemicals by using a mild detergent or soap (preferred) and water; do not use neutralizing chemicals, unguents, or salves.

Chemicals on the Skin in a Confined Area

Immediately flush with cold water and wash by using a mild detergent or soap (preferred) and water. If there is no visible burn, scrub with warm water and soap, removing any jewelry in the affected area. If a delayed action [the physiological effects of some chemicals (e.g., methyl and ethyl bromides) may be delayed as much as 48 hours] is noted, obtain medical attention promptly and explain carefully what chemicals were involved.

I.G

Design Requirements for and Use of Electrically Powered Laboratory Apparatus

During the past 25 years, the use of electrically powered apparatus in laboratories has increased more rapidly than that of any other category of equipment. Such equipment is now used routinely for operations requiring heating, cooling, agitation or mixing, and pumping, as well as for a variety of instruments used in making physical measurements. In fact, electrical apparatus is now so commonly available that it, rather than burners or other devices that have open flames, should be the only type of heat source present in operations involving the use of flammable materials. However, although the introduction of electrically powered equipment has resulted in a major improvement in laboratory safety, the use of this equipment does pose a new set of possible hazards for the unwary.

I.G.1 GENERAL PRINCIPLES

Laboratory workers should know the procedures for removing a person from contact with a live electrical conductor and the emergency first aid procedures to use for a person who has received a serious electrical shock (see Section I.F.6).

All 110-V outlet receptacles in laboratories should be of the standard design that accepts a three-prong plug and provides a ground connection. The use of an old-style two-prong receptacle and an adapter that takes a three-prong plug is a less satisfactory alternative; however, no attempt should be made to bypass the ground. Old-style receptacles should be replaced as soon as feasible, and an additional ground wire should be

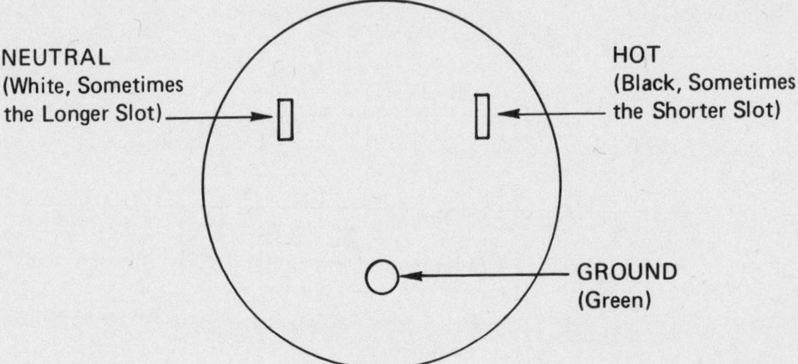

FIGURE 4 Standard wiring connection for 110-V receptacle: front view.

added if necessary so that each receptacle is wired as shown in Figure 4. If the use of an extension cord becomes necessary, standard three-conductor extension cords that provide an independent ground connection should be used.

It is also possible to fit a receptacle with a ground-fault circuit breaker that will disconnect the current if continuity is lost in the system. Such ground-fault protection devices are frequently recommended for outdoor receptacles by local electrical codes and would be useful for selected laboratory receptacles where maintenance of a good ground connection is essential for safe operation.

Receptacles that provide electric power for operations in hoods should be located outside of the hood. This location prevents the production of electrical sparks inside the hood when a device is plugged in and also allows a laboratory worker to disconnect electrical devices from outside the hood. However, cords should not be allowed to dangle outside the hood in such a way that a worker could accidently pull them.

Laboratory equipment that is to be plugged into a 110-V (or higher) receptacle should be fitted with a standard three-conductor line cord that provides an independent ground connection to the chassis of the apparatus (see Figure 5) (although in some instances "double-insulated" equipment that has a two-conductor line cord may be adequate). All frayed or damaged line cords should be replaced before further use of the equipment is permitted; annual inspection of all electrical cords is good practice. It is also desirable that equipment that will be plugged into an electrical receptacle be fitted with a fuse or other overload-protection device that will disconnect the electrical circuit in the event the apparatus fails or is overloaded. This overload protection is particularly useful for apparatus

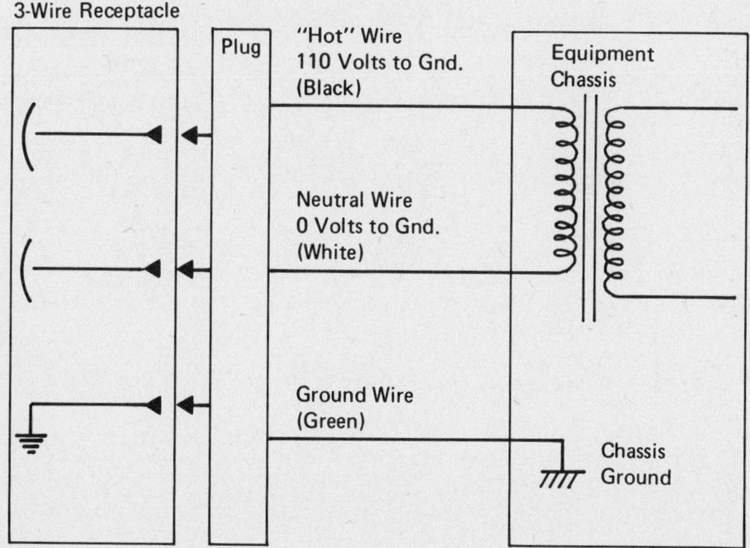

FIGURE 5 Standard wiring convention for line cord supplying 110-V electrical power to equipment.

likely to be left on and unattended for long time periods [such as variable autotransformers (Variacs® and Powerstats®), vacuum pumps, drying ovens, stirring motors, and electronic instruments]. New or existing equipment that does not contain this overload protection can be modified to provide such protection.

Motor-driven electrical equipment used in a laboratory where volatile flammable materials may be present should be equipped with a nonsparking induction motor rather than with a series-wound motor that uses carbon brushes. This applies to the motors used in vacuum pumps, mechanical shakers, and (especially) stirring motors, magnetic stirrers, and rotary evaporators. The speed of an induction motor operating under a load should not be controlled by using a variable autotransformer; such circumstances will cause the motor to overheat and might start a fire. There is no way to modify an apparatus that has a series-wound motor so that it will be spark-free. For this reason, many kitchen appliances (mixers, blenders, and such) should not be used in laboratories where flammable materials may be present. Finally, it should be remembered that, when other items of equipment (especially vacuum cleaners and portable electric drills) having series-wound motors are brought into a laboratory for special purposes, special precautions (see Section I.C.1) should be taken to ensure that no flammable vapors are present before such equipment is used.

Whenever possible, electrical equipment should be located so as to minimize the possibility that water or chemicals could accidentally be spilled on it. If water or any chemical should accidentally be spilled on electrical equipment, the equipment should be unplugged immediately and not used again until it has been cleaned and inspected (preferably by a qualified technician). Water can also enter electrical equipment from condensation on equipment placed in a cold room or a large refrigerator. Cold rooms pose a particular hazard in this respect (see Section I.H.4). The atmosphere in such rooms is frequently at a high relative humidity and the potential for water condensate is significant. If electrical equipment must be placed in such areas, the condensation problem can be lessened (but not eliminated) by mounting the equipment on a wall or vertical panel. The potential for electrical shock in these rooms can be minimized by careful electrical grounding of the equipment and the use of a suitable flooring material.

All laboratories should have access to a qualified technician who can make routine repairs to existing equipment and modifications to new or existing equipment so that it will meet reasonable standards for electrical safety [see the current volume of the *National Electrical Code* (National Fire Protection Association, Boston, Mass.)].

With the exception of certain instrument adjustments (see Section I.G.7), line cords of electrical equipment should always be unplugged before any adjustments, modifications, or repairs are undertaken. When it is necessary to handle a piece of electrical equipment that is plugged in, laboratory workers should first be certain that their hands are dry.

I.G.2 VACUUM PUMPS

Distillations or concentration operations that involve significant quantities of volatile substances should normally be performed by using a water aspirator or a steam aspirator, rather than by using a mechanical vacuum pump. However, the distillation of less-volatile substances, removal of final traces of solvents, and some other operations require pressures lower than those that can be obtained by using a water aspirator and are normally performed by using a mechanical vacuum pump (see Section I.D.3). The input line from the system to the vacuum pump should be fitted with a cold trap to collect volatile substances from the system and minimize the amount that enters the vacuum pump and dissolves in the pump oil; the use of liquid nitrogen or liquid air in such traps, however, can lead to a flammability hazard (see Section I.C.1). The possibility that mercury will be swept into the pump as a result of a sudden loss of vacuum can be minimized by placing a Kjeldahl trap in the line to the pump.

The output of each pump should be vented to an air exhaust system (see Section I.H.3). This procedure is essential when the pump is being used to evacuate a system containing a volatile toxic or corrosive substance (failure to observe this precaution would result in pumping the substance into the laboratory atmosphere); it may also be necessary to scrub or absorb vapors. Even with these precautions, however, volatile toxic or corrosive substances may accumulate in the pump oil and, thus, be discharged into the laboratory atmosphere during future pump use. This hazard should be avoided by draining and replacing the pump oil when it becomes contaminated. The contaminated pump oil should be disposed of by following standard procedures for the safe disposal of toxic or corrosive substances (see Chapters II.E and G).

Belt-driven mechanical pumps having exposed belts should have protective guards. Such guards are particularly important for pumps installed on portable carts or tops of benches where laboratory workers might accidentally entangle clothing or fingers in the moving belt, but are not necessary for pumps located at a height of at least 7 ft above the working surface or in an enclosed cabinet.

I.G.3 DRYING OVENS

Electrically heated ovens are commonly used in the laboratory to remove water or other solvents from chemical samples and to dry laboratory glassware before its use. With the exception of vacuum drying ovens, these ovens rarely have any provision for preventing the discharge of the substances volatilized in them into the laboratory atmosphere. Thus, it should be assumed that these substances will escape into the laboratory atmosphere and could also be present in concentrations sufficient to form explosive mixtures with the air inside the oven (see Section I.C.1).

Ovens should not be used to dry any chemical sample that has even moderate volatility and might pose a hazard because of acute or chronic toxicity unless special precautions have been taken to ensure continuous venting of the atmosphere inside the oven. Thus, most organic compounds should not be dried in a conventional laboratory oven.

Glassware that has been rinsed with an organic solvent should not be dried in an oven. If such rinsing is necessary, the item should be rinsed again with distilled water before being placed in the oven.

Because of the possible formation of explosive mixtures by volatile substances and the air inside an oven, laboratory ovens should be constructed so that their heating elements (which may become red hot) and their temperature controls (which may produce sparks) are physically separated from their interior atmospheres. Many small household ovens and similar heating devices do not meet these requirements and, consequently, should not be used in laboratories. Existing ovens that do not meet these requirements should either be modified or have a sign attached to the oven door to warn workers that flammable materials should not be placed in that oven. Some safety groups suggest that every laboratory oven should be modified by placing a blow-out panel in its rear wall so that any explosion within the oven will not blow the oven door and the oven contents into the laboratory. NFPA standards call for blow-out vents on ovens handling flammable substances.

Thermometers containing mercury should not be mounted through holes in the tops of ovens so that the bulb of mercury hangs into the oven. Bimetallic strip thermometers are the preferred alternative for monitoring oven temperatures. Should a mercury thermometer be broken in an oven of any type, the oven should be turned off and all mercury should be removed from the cold oven.

I.G.4 REFRIGERATORS

The potential hazards posed by laboratory refrigerators are in many ways similar to those of laboratory drying ovens. Because there is almost never a satisfactory arrangement for continuously venting the interior atmosphere of a refrigerator, any vapors escaping from vessels placed in one will accumulate. Thus, the atmosphere in a refrigerator could contain an explosive mixture of air and the vapor of a flammable substance or a dangerously high concentration of the vapor of a toxic substance or both. (The problem of toxicity is aggravated by the practice of laboratory workers who place their faces inside the refrigerator while searching for a particular sample, thus ensuring the inhalation of some of the atmosphere from the refrigerator interior.) As noted in Section I.A.3, laboratory refrigerators should never be used for the storage of food or beverages.

There should be no potential sources of electrical sparks on the inside of a laboratory refrigerator. In general, it is preferable when purchasing a refrigerator for a laboratory to select a "flammable storage" one that has been so designed by the manufacturer. If this is not possible, all new or existing refrigerators should be modified as needed by (a) removing any interior light activated by a switch mounted on the door frame, (b) moving the contacts of the thermostat controlling the temperature to a position outside the refrigerated compartment, and (c) moving the contacts for any thermostat present to control fans within the refrigerated compartment to the outside of the refrigerated compartment. Although a prominent sign warning against the storage of flammable substances could be permanently attached to the door of an unmodified refrigerator, this alternative is less desirable than the modification of the equipment by removal of all spark sources from the refrigerated compartment. "Frost-free" refrigerators are not advisable for laboratory use because of the many problems associated with attempts to modify them. Many of these refrigerators have a drain tube or hole that carries water (and any flammable material present) to an area adjacent to the compressor and, thus, present a spark hazard; the electric heaters used to defrost the freezing coils are also a potential spark hazard (see Section I.C.1).

Laboratory refrigerators should be placed against fire-resistant walls, have heavy-duty cords, and preferably should be protected by their own circuit breaker.

Uncapped containers of chemicals should never be placed in a refrigerator. Containers of chemicals should be capped so as to achieve a seal that is both vapor tight and unlikely to permit a spill if the container is tipped over. Caps constructed from aluminum foil, corks, corks wrapped with aluminum foil, or glass stoppers often do not meet all of these criteria; the use of such methods for capping containers should be discouraged. The most satisfactory temporary seals are normally achieved by using containers that have screw-caps lined with either a conical polyethylene insert or a Teflon® insert. The best containers for samples that are to be stored for longer periods of time are sealed, nitrogen-filled glass ampoules.

The placement of potentially explosive (see Section 2) or highly toxic substances (see Chapter I.B) in a laboratory refrigerator is strongly discouraged. If this precaution must be violated, then a clear, prominent warning sign should be placed on the outside of the refrigerator door. The length of storage of such material in the refrigerator should be kept to a minimum.

I.G.5 STIRRING AND MIXING DEVICES

The stirring and mixing devices commonly found in laboratories include stirring motors, magnetic stirrers, shakers, small pumps for fluids, and rotary evaporators for solvent removal. These devices are typically used in laboratory operations that are performed in a hood (see Section I.H.1), and it is important that they be operated in a way that precludes the production of electrical sparks within the hood (see Section I.G.1). Furthermore, it is important that, in the event of an emergency situation, a laboratory worker be able to turn such devices on or off from outside the hood. Finally, heating baths associated with these devices (e.g., baths for rotary evaporators) should also be spark-free and capable of control from outside the hood.

Only spark-free induction motors should be used to run stirring and mixing devices; any motor that may produce sparks during its start-up or running cycle should not be used. Fortunately, the motors in most of the currently marketed stirring and mixing devices meet this criterion. However, the on-off switches and rheostat-type speed controls of many of them do not meet this criterion because the switch or rheostat has exposed contacts that can produce an electrical spark any time a change is made in the controls. This problem is particularly true of many of the magnetic stirrers and rotary evaporators currently being sold. An effective solution is the modification of the stirring or mixing device by removing any switches located on the device and inserting a switch in the line cord near the plug end. As the electrical receptacle for the plug should be outside the hood, this modification will ensure that the switch will also be outside of the hood. Recall that the speed of an induction motor operating under a load should not be controlled by a variable autotransformer (see Section I.G.1).

Because stirring and mixing devices (especially stirring motors and magnetic stirrers) are often operated for fairly long periods without continual attention (e.g., reaction mixtures that are stirred overnight), the consequences of stirrer failure, electrical overload, or blockage of the motion of the stirring impeller should be considered. It is good practice to attach a stirring impeller to the shaft of the stirring motor by using lightweight rubber tubing. Then, if the motion of the impeller becomes blocked (e.g., by formation of a copious precipitate), the rubber simply twists until it breaks rather than either the motor stalling or the glass apparatus into which the stirring impeller extends breaking. It is also very desirable (but unfortunately not very common) that stirring motors that are to be left unattended be fitted with a suitable fuse or thermal-protection device (see Section I.G.1).

I.G.6 HEATING DEVICES

Perhaps the most common electrical equipment found in a laboratory are the devices used to supply the heat needed to effect a reaction or a separation. (The use of steam-heated devices rather than electrically heated devices is generally preferred whenever temperatures of 100°C or less are required; these devices do not present shock or spark hazards and can be left unattended with assurance that their temperature will never exceed 100°C.) These electrically heated devices include hot plates, heating mantles and tapes, oil baths, air baths, hot-tube furnaces, and hot-air guns. Although they are inherently much safer than burners as laboratory heat sources, such devices can still pose both electrical and fire hazards if used improperly.

1. The actual heating element in any laboratory heating device should be enclosed in a glass, ceramic, or insulated metal case such that it is not possible for the laboratory worker (or some metallic conductor) to accidentally touch the wire carrying the electric current. This practice minimizes the hazards of electrical shock and of accidentally producing an electrical spark near a flammable liquid or vapor (see Section I.C.1). This type of construction also diminishes the possibility that a flammable liquid or vapor will come in contact with the hot wire (whose temperature is frequently higher than the ignition temperature of many common solvents). If any heating device becomes so worn or damaged that its heating element is exposed, the device should either be discarded or repaired to correct the damage before it is again used in the laboratory. Note that many household appliances (e.g., hot plates and space heaters) do not meet this criterion and, consequently, are not advisable for use in the laboratory.

2. The temperature of many laboratory heating devices (e.g., heating mantles, air baths, and oil baths) is controlled by use of a variable autotransformer that supplies some fraction of the total line voltage (typically 110 V) to the heating element of the device. New or existing variable autotransformers should be wired (or rewired) as illustrated in Figure 6. If a variable autotransformer is not wired in this manner (older models were not), the switch on it may or may not disconnect both wires of the output from the 110-V line when in the off position. Also, if this wiring scheme has not been followed (especially the use of a grounded three-prong plug), even when the potential difference between the two output lines is only 10 V, each output line may be at a relatively high voltage (e.g., 110 and 100 V) with respect to an electrical ground. Because of these possibilities, whenever a variable autotransformer whose wiring is not definitely known to be acceptable is used, it is best to assume that either of the output lines could be at a potential of 110 V and capable of delivering a lethal electric shock. The cases of all variable autotransformers have numerous openings to allow for ventilation and some sparking may occur whenever the voltage adjustment knob is turned; laboratory workers should be careful to locate these devices where water and other chemicals cannot be spilled on them (shock hazard) and where their movable contacts will not be exposed to flammable liquids or vapors (fire hazard). Specifically, variable autotransformers should be mounted on

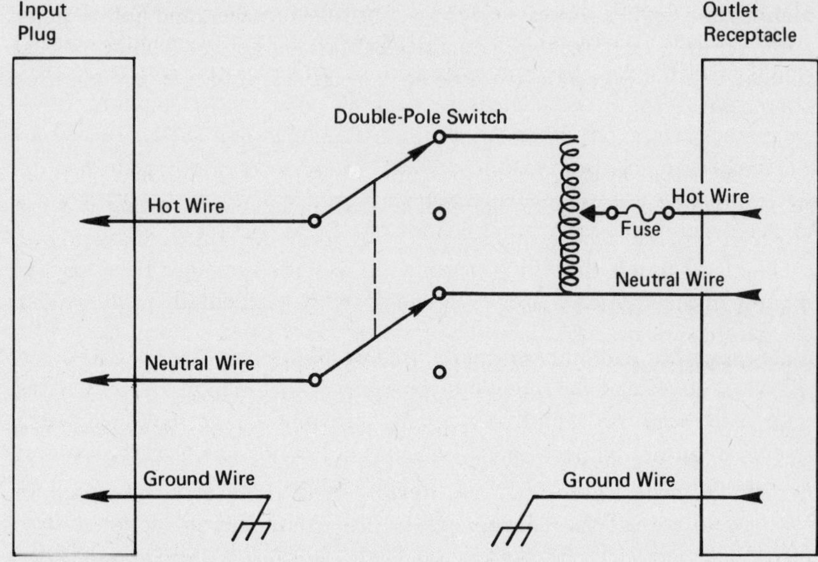

FIGURE 6 Schematic diagram of properly wired variable autotransformer.

walls or vertical panels and outside of hoods; they should not just be placed on laboratory bench tops, especially those inside of hoods.

3. Because the electrical input lines (even lines from variable transformers) to almost all laboratory heating devices may well be at a potential of 110 V with respect to any electrical ground, these lines should always be considered both as potential shock hazards and as potential spark hazards (see Section I.C.1). Thus, any connection from these lines to a heating device should be both mechanically and electrically secure and completely covered with some insulating material. Alligator clips should not be used to connect a line cord from a variable autotransformer to a heating device (especially an oil bath or an air bath) because such connections pose a shock hazard and also may slip off, creating an electrical spark and, perhaps, contacting other metal parts to create a new hazard. All connections should be made by using either insulated binding posts or, preferably, a plug and receptacle combination.

4. Whenever an electrical heating device is to be left unattended for a significant period of time (e.g., overnight), it is advisable that it be equipped with a temperature-sensing device that will turn off the electric power if the temperature of the heating device exceeds some preset limit. Similar control devices are available that will turn off the electric power if

the flow of cooling water through a condenser is unexpectedly stopped. Such fail-safe devices, which can either be purchased or constructed by a qualified technician, prevent more serious problems (fires or explosions) that may arise if the temperature of an unattended reaction should increase significantly either because of a change in line voltage or because of accidental loss of reaction solvent. These devices are also valuable accessories for use with stills employed to purify reaction solvents because such stills are often left unattended for significant periods of time.

Laboratory Hot Plates

Laboratory hot plates are normally used when solutions are to be heated to 100°C or above and the inherently safer steam baths cannot be used as the source of heat. As noted above, only hot plates that have completely enclosed heating elements should be used in laboratories. Although almost all laboratory hot plates now being sold meet this criterion, many older ones pose an electrical-spark hazard arising from the on-off switch located on the hot plate, the bimetallic thermostat used to regulate the temperature of the hot plate, or both. Normally, these two spark sources are both located in the lower part of the hot plate in a region where any heavier-than-air (and possibly flammable) vapor escaping from a boiling liquid on the hot plate would tend to accumulate. In principle, these spark hazards could be alleviated by enclosing all mechanical contacts in a sealed container or by the use of solid-state circuitry for switching and temperature control but, in practice, such modifications would be difficult to incorporate in many of the hot plates now used in laboratories. Laboratory workers should be warned of the spark hazard associated with many existing hot plates (perhaps by attaching a warning label), and any new hot plates purchased should be constructed in a way that avoids electrical spark hazards. In addition to the spark hazard, old and corroded bimetallic thermostats in these devices can eventually fuse shut, thus delivering full current to the hot plate. This hazard can be avoided by wiring a fusible coupling into the line inside the hot plate so that if the device does overheat, the coupling will melt and interrupt the current (see Section I.G.1).

Heating Mantles

Heating mantles are commonly used for heating round-bottomed flasks, reaction kettles, and related reaction vessels. These mantles are constructed by enclosing the heating element in a series of layers of fiberglass cloth. As long as the fiberglass coating is not worn or broken or no water or other chemicals are spilled into the mantle (see Section I.G.1), these mantles pose no shock hazard to the worker. They are normally fitted with a *male* plug that fits into a *female* receptacle on an output line from a variable autotransformer to provide a connection that is mechanically and electrically secure.

Heating mantles are always intended to be used with a variable autotransformer to control the input voltage and should never be plugged directly into a 110-V line. Laboratory workers should be careful not to exceed the input voltage recommended by the manufacturer of the mantle because higher voltages will cause it to overheat, melting the fiberglass insulation and exposing the bare (and often red-hot) heating element. Note that the maximum recommended input voltage for a mantle that is being used with a dry flask is 10-20 V lower than that for a mantle that is being used with a flask containing a liquid. Some heating mantles are constructed by encasing the fiberglass mantle in an outer metal case that provides physical protection against damage to the fiberglass. If such metal-enclosed mantles are used, it is good practice to ground the outer metal case either by use of a three-conductor cord (containing a ground wire) from the variable autotransformer or by attaching one end of a heavy braided conductor to the mantle case and the other end to a good electrical ground such as a cold-water pipe. This practice provides the laboratory worker with protection against an electric shock if the heating element inside the mantle should be shorted against the metal case.

Oil Baths

Electrically heated oil baths are often used as heating devices for small or irregularly shaped vessels or when a stable heat source that can be maintained at a constant temperature is desired. For temperatures below 200°C, a saturated paraffin oil is often used; a silicone oil (which is more expensive) should be used for temperatures up to 300°C. An oil bath should always be monitored by using a thermometer or other thermal-sensing device to ensure that its temperature does not exceed the flash point of the oil being used. For the same reason, oil baths left unattended should be fitted with thermal-sensing devices that will turn off the electric power if the bath overheats. Bare wires should not be used as resistance devices to heat oil baths. Rather, these baths should be heated with a metal pan fitted with an enclosed heating element or with a heating element enclosed in a metal sheath (i.e., a knife heater, a tubular immersion heater such as a Calrod®, or its equivalent). The input connection for this heating element should be a *male* plug that will fit a *female* receptacle from a variable autotransformer output line.

Heated oil should be contained in either a metal pan or a heavy-walled porcelain dish; a glass dish or beaker can break and spill hot oil if accidentally struck with a hard object. The oil bath should be carefully mounted on a stable horizontal support such as a laboratory jack that can be easily raised or lowered without danger of the bath tipping over. A bath should never be supported on an iron ring because of the greater likelihood of accidentally tipping it over. Finally, a laboratory worker using an oil bath heated above 100°C should be careful to guard against the possibility that water (or some other volatile substance) could fall into the hot bath. Such an accident can splatter hot oil over a wide area.

AIR BATHS

Electrically heated air baths are frequently used as a substitute for heating mantles when heating small or irregularly shaped vessels. Because of their inherently low heat capacity, such baths normally must be heated considerably above (100°C or more) the desired temperature of the vessel being heated. These baths should be constructed so that the heating element is completely enclosed and the connection to the air bath from the variable transformer is both mechanically and electrically secure. These baths can be constructed from metal, ceramic, or (less desirably) glass vessels. If glass vessels are used, they should be thoroughly wrapped with a heat-resistant tape so that, if the vessel is accidentally broken, the glass will be retained and the bare heating element will not be exposed.

HEAT GUNS

Laboratory heat guns are constructed with a motor-driven fan that blows air over an electrically heated filament. They are frequently used to dry glassware or to heat the upper parts of a distillation apparatus during distillation of high-boiling materials. The heating element in a heat gun typically becomes red hot during use and, necessarily, cannot be enclosed. Also, the on-off switches and fan motors are not usually spark free. For these reasons, heat guns almost always pose a serious spark hazard (see Section I.C.1). They should never be used near open containers of flammable liquids, in environments where appreciable concentrations of flammable vapors may be present, or in hoods that are being used to remove flammable vapors. Household hair dryers should be used as substitutes for laboratory heat guns only if they have three-conductor or double-insulated line cords.

I.G.7 ELECTRONIC INSTRUMENTS

Most modern electronic instruments are fitted with a line cord that contains a separate ground wire for the chassis and with a suitable fuse or other overload protection. Any existing instrument that lacks these features should be modified to incorporate them. As with any other electrical equipment, special precautions should be taken to avoid the possibility that water or other chemicals could be spilled on these instruments (see Section I.G.1).

Under most circumstances, any repairs to, adjustments of, or alterations in such instruments should be made by a qualified technician. If laboratory workers do undertake repairs, the instrument line cord should always be unplugged before any disassembly is begun. Certain instrument adjustments can be made only when the instrument is connected to the power. Laboratory workers should not undertake such adjustments without supervision unless they have received specific prior instruction. This precaution is particularly important with instruments that incorporate high-voltage circuitry (such as oscilloscopes, spectrometers that have photomultiplier tubes, and most equipment that uses vacuum-tube circuitry).

I.H
Laboratory Ventilation

Laboratories that use chemicals vary from spacious well-designed facilities to those that consist of a single room that has been designated as a laboratory and has little or no provision for ventilation. The need for ventilation of these different types of laboratories will vary from the provision of simple comfort for the occupants to the control of highly toxic volatile substances. This discussion will center on ventilation for the control of toxic chemicals; however, the overall performance of laboratory workers will also benefit from ventilation systems that control the temperature, humidity, and concentration of odoriferous materials in the laboratory.

There is a tendency for laboratory workers to associate odor with toxicity. This tendency may result in overconcern for an odoriferous substance of low toxicity and a lack of concern for highly toxic substances that have little or no odor.

The steady increase in the cost of energy in recent years has resulted in a conflict between the desire to minimize the costs of heating or cooling and dehumidifying laboratory air and the need to provide laboratory workers with improved ventilation as a means of protection from toxic gases, vapors, aerosols, and dusts. However, although the energy costs associated with tempering the input air for laboratories are often substantial, cost considerations should never take precedence over ensuring that laboratories have adequate ventilation systems to protect workers from hazardous concentrations of airborne toxic substances. Thus, any changes in overall laboratory ventilation systems to conserve energy should be instituted only after thorough testing of their effects has demonstrated that laboratory workers will still have adequate protection. An inadequate ventilation system can be worse than none, because it is likely to give laboratory workers an unwarranted sense of security that they are protected from airborne toxic substances.

I.H.1 GENERAL LABORATORY VENTILATION

General ventilation refers to the quantity and quality of the air supplied to the laboratory. The overall ventilation system should ensure that the laboratory air is continuously being replaced so that concentrations of odoriferous or toxic substances do not increase during the working day. Provided that auxiliary local exhaust systems (see Sections I.H.2 and 3) are available and are used as the primary method for controlling concentrations of airborne substances, a ventilation system that changes the room air 4-12 times per hour is normally adequate.

In all cases, the movement of air in the general ventilation system for a building should be from the offices, corridors, and such into the laboratories. All air from laboratories should be exhausted outdoors and not recycled. Thus, the air pressure in the laboratories should always be negative with respect to the rest of the building. The air intakes for a laboratory building should be in a location that reduces the possibility that the input air will be contaminated by the exhaust air from either the same building or any other nearby laboratory building. One common arrangement is to locate all of the laboratory-hood exhaust vents (the usual exhaust ports for laboratories) on the roof of the laboratory building and the building air-intake port at a different site where local movement is unlikely to mix exhaust air and intake air.

Laminar (i.e., nonturbulent) flow of incoming air is ideal and can be approached by using a plenum or several louvers at the air-input sites of the laboratory. Air entry through perforated ceiling panels has also been used successfully to provide a uniform airflow. The plenum, louver, or perforated ceiling panels should be designed so as to direct the clean, incoming air over the laboratory personnel and sweep the contaminated air away from their breathing zone.

The size of the room and its geometry or configuration as well as the velocity and volume of input air will affect the room air patterns. However, it is difficult to offer generalizations about the effects of air-input and air-output ports on general laboratory ventilation.

Evaluation of General Laboratory Ventilation

Each laboratory should be evaluated for the quality and quantity of general ventilation present. This evaluation should be repeated periodically and any time a change is made, either in the general ventilation system for the building or in some aspect of the local ventilation within the laboratory. This evaluation should begin by observing the pattern of air movement entering and within the laboratory.

Airflow paths into and within a room can be determined by observing smoke patterns. Convenient smoke sources are commercial smoke tubes available from local safety and laboratory supply houses (see Section I.H.7). Aerosol generators that produce continuous, voluminous fogs are also available from the same sources. Such aerosol generators are generally used in ventilation research and for evaluating local exhaust ventilation, such as laboratory hoods. If the general laboratory ventilation is satisfactory, the movement of air from the corridors and other input ports through the laboratory to the hoods or other exhaust ports should be relatively uniform. There should be no areas where the air remains static or areas that have unusually high airflow velocities. If areas that have little or no air movement are found, a ventilation engineer should be consulted and appropriate changes should be made in input or output ports to correct the deficiencies. Alternatively, signs warning of inadequate ventilation should be posted in such areas.

At low concentrations, most chemical vapors and gases tend to rise with the warm air currents and become diluted with the general room air. Air movement in large laboratories is normally multidirectional and typically has a velocity of about 20 linear ft/min. This diverse air movement is a result of the movement of people and the effects of air intakes and exhausts and of eddy currents around benches and other fixed objects. The net result is that the general air composition is rather uniform. This mixing of room air does not mean that isolated static air spaces cannot be found, but rather that the general area occupied by the laboratory workers tends to be uniform in composition unless there are serious deficiencies in the locations of input and output ports.

The average time required for a ventilation system to change the air within a laboratory can be estimated from the total volume of the laboratory (usually measured in cubic feet) and the rate at which input air is introduced or exhaust air is removed [usually measured in cubic feet per minute (cfm)]. The latter value is usually determined by measuring [usually in linear feet per minute (lfm)] the average face velocity for each laboratory exhaust port, such as the hoods (see Section I.H.2) or other local ventilation systems. For each exhaust port, the product of the face area (in square feet) and the average face velocity (in lfm) will give the rate at which air is being exhausted by that port (in cfm). The sum of these rates for all exhaust ports in the laboratory will give the total rate at which air is being exhausted from the laboratory. It is important to realize that, up to the capacity of the exhaust system, the rate at which air is exhausted from the laboratory will equal the rate at which input air is introduced. Thus, decreasing the flow rate of input air (perhaps to conserve energy) will decrease the number of air changes per hour in the laboratory, the face velocities of the hoods, and the capture velocities of all other local ventilation systems.

The measurement of airflow rates requires special instruments and personnel trained to use them. Pitot tubes are used for measuring duct velocities, and anemometers or velometers are used to measure airflow rates within rooms and at the faces of input or exhaust ports. These instruments are available from safety supply companies or laboratory supply houses (see Section I.H.6). The proper calibration and use of these instruments and the evaluation of the data are a separate discipline; consultation with an industrial hygienist or a ventilation engineer is recommended whenever serious ventilation problems are suspected or when decisions on appropriate changes in the ventilation system to achieve a proper balance of input and exhaust air must be made.

Evaluation of Airborne Contaminants by Using TLVs

It is possible to calculate the amount of a chemical that can be emitted into the general laboratory atmosphere without exceeding the TLV or acceptable-exposure value. The air-saturation level (ASL) (in parts per million) can be calculated from the vapor pressure (P)(in mm Hg) by using Equation 1:

$$\text{ASL} = (10^6\text{P})/760. \quad (1)$$

The results for some typical chemicals are given in Table 5. Most of these saturation levels are greater than the established TLVs, and some may even be life threatening. For this reason, protective apparel including self-contained breathing apparatus (see Section I.F.4) should be used to clean up major spills (see Section II.E.6) of highly toxic volatile chemicals because concentrations approaching saturation may be possible. Such chemical spills may also lead to airborne concentrations of chemicals in the explosive range (see Chapter I.C), so that care should be taken to avoid any ignition source during cleanup operations.

Another illustrative calculation that can be made is the amount of a

TABLE 5 Air-Saturation Concentration of Common Solvents

Solvent	Vapor Pressure at 20°C (mm Hg)	Saturation Level at 20°C (ppm)	TLV (ppm)
Acetone	184.8	2.43×10^5	1000
Benzene	74.2	9.76×10^4	10
Carbon tetrachloride	92.0	1.21×10^5	10
Chloroform	160.0	2.11×10^5	10
Diethyl ether	430.0	5.66×10^5	400
Dioxane	30.0	3.95×10^4	50
Ethanol	43.0	5.66×10^4	1000
Ethylene glycol	0.05	65.8	100
Hexane	119.0	1.57×10^5	100
Methylchloroform	100.0	1.32×10^5	350
Methylene chloride	349.0	4.59×10^5	200
Toluene	22.0	2.89×10^4	100

chemical that can be volatilized into the general laboratory atmosphere without exceeding the TLV, assuming a static atmosphere in the laboratory. For W grams of a substance of molecular weight M and a laboratory that has a volume V (in cubic feet), the concentration (C) (in ppm) at 25° is given by Equation 2:

$$C = 8.65 \times 10^5(W)/(V)(M). \tag{2}$$

Thus, if 454 g (1 lb) of acetone is spilled and allowed to evaporate in a laboratory that has a volume of 1000 ft^3, the concentration of acetone will be 6770 ppm, which substantially exceeds the TLV of 1000. The maximum amount of acetone that can be released into a laboratory of this size without exceeding the TLV is 67 g. For a more toxic material, such as benzene (TLV = 10 ppm), the maximum permissible amount of material that can be released into a laboratory having a volume of 1000 ft^3 is about 1 g. The corresponding values for several common solvents are given in Table 6.

Such calculations illustrate the basis for the ACGIH recommendation that auxiliary local ventilation be used when working with a substance having a TLV of 50 ppm or less [See Committee on Industrial Ventilation (Section I.H.7)]. For many such substances, 5 g or less per 1000 ft^3 of laboratory atmosphere will exceed the TLV.

More-accurate estimations of general-laboratory airborne concentrations of chemicals can be made if the dilution resulting from continuous air change is also considered. By assuming rapid diffusion and mixing, the general airborne concentration of a substance can be calculated by

TABLE 6 Amount of Vaporized Solvent per 1000 ft^3 of Air Required to Equal the TLV at 25°C

Solvent	*M*r	TLV (ppm)	Limit (g)
Acetone	58	1000	67
Benzene	78	10	1
Carbon tetrachloride	154	10	2
Chloroform	119	10	1
Diethyl ether	74	400	34
Dioxane	88	50	5
Ethanol	46	1000	53
Ethylene glycol	62	100	7
Hexane	86	100	10
Methylene chloride	85	200	20
Toluene	92	100	11
1,1,1-Trichloroethane	133	350	55

estimating its total emission into the room and dividing this by the rate of air exhaust from the room. For typical laboratories where work with a variety of different substances is in progress, additional calculations of this type are normally not warranted. If a reliable measure of the exposure to a specific substance is required, it is better obtained by having a laboratory worker wear a portable air-sampling device, packed with a suitable adsorbent and mounted in the breathing zone, for a fixed time period. Analysis of the content of the sampling device provides a more accurate measure of the actual concentration of a specific substance to which the worker has been exposed. Suitable air-sampling devices are available from various safety supply companies and laboratory supply houses (see Section I.H.6).

USE OF GENERAL LABORATORY VENTILATION

As discussed above, general laboratory ventilation is intended primarily to increase the comfort of laboratory workers and to provide a supply of air that will be exhausted by a variety of auxiliary local ventilation devices (hoods, vented canopies, vented storage cabinets, and such). This ventilation provides only very modest protection from toxic gases, vapors, aerosols, and dusts, especially if they are released into the laboratory in any significant quantity. The cardinal rule for safety in working with toxic substances (see Chapter I.B) is that all work with these materials in a laboratory should be performed in such a way that they do not come in contact with the skin and that quantities of their vapors or dust that might produce adverse toxic effects are prevented from entering the general laboratory atmosphere. Thus, operations such as running reactions, heating or evaporating solvents, and transfer of chemicals from one container to another should normally be performed in a hood. If especially toxic or corrosive vapors will be evolved, these exit gases should be passed through scrubbers or adsorption trains. Toxic substances should be stored in cabinets fitted with auxiliary local ventilation (see Chapters II.B and D), and laboratory apparatus that may discharge toxic vapors (vacuum pump exhausts, gas chromatograph exit ports, liquid chromatographs, distillation columns, and such) should be vented to an auxiliary local exhaust system such as a canopy or a snorkel. Samples that will be measured by using instruments or stored in apparatus where auxiliary local ventilation is not practical [such as balances, spectrometers, and refrigerators (see Section I.G.4)] should be kept in closed containers during measurement or storage. Simply stated, laboratory workers should regard the general laboratory atmosphere only as a source of air to breathe and as a source of input air for auxiliary local ventilation systems.

I.H.2 USE OF LABORATORY HOODS

Although many laboratory workers regard hoods strictly as local ventilation devices to be used to prevent toxic, offensive, or flammable vapors from entering the general laboratory atmosphere, hoods offer two other significant types of protection. Placing a reacting chemical system within a hood, especially with the hood sash closed, also places a physical barrier between the workers in the laboratory and the chemical reaction. This barrier can afford the laboratory workers significant protection from hazards such as chemical splashes or sprays, fires, and minor explosions. Furthermore, the hood can provide an effective containment device for accidental spills of chemicals. In a laboratory where workers spend most of their time working with chemicals, there should be at least one hood for each two workers, and the hoods should be large enough to provide each worker with at least 2.5 linear ft of working space at the face. The optimum arrangement is to provide each laboratory worker with a separate hood. In circumstances where this amount of hood space cannot be provided, there should be reasonable provisions for other types of local ventilation and special care should be exercised in monitoring and restricting the use of hazardous substances.

The following factors should be remembered in the daily use of hoods:

1. Hoods should be considered as backup safety devices that can contain and exhaust toxic, offensive, or flammable materials when the

design of an experiment fails and vapors or dusts escape from the apparatus being used. Hoods should not be regarded as means for disposing of chemicals. Thus, apparatus used in hoods should be fitted with condensers, traps, or scrubbers to contain and collect waste solvents or toxic vapors or dusts. Highly toxic or offensive vapors should always be scrubbed or adsorbed before the exit gases are released into the hood exhaust system.

2. Hoods should be evaluated before use to ensure adequate face velocities (typically 60-100 lfm) [see Fuller and Etchells (Section I.H.7)] and the absence of excessive turbulence (see below). Further, some continuous monitoring device for adequate hood performance should be present and should be checked before each hood is used. If inadequate hood performance is suspected, it should be established that the hood is performing adequately before it is used.

3. Except when adjustments of apparatus within the hood are being made, the hood should be kept closed: vertical sashes down and horizontal sashes closed. Sliding sashes should not be removed from horizontal sliding-sash hoods. Keeping the face opening of the hood small improves the overall performance of the hood.

4. The airflow pattern, and thus the performance of a hood, depends on such factors as placement of equipment in the hood, room drafts from open doors or windows, persons walking by, or even the presence of the user in front of the hood. For example, the placement of equipment in the hood can have a dramatic effect on its performance. Moving an apparatus 5-10 cm back from the front edge into the hood can reduce the vapor concentration at the face by 90%.

5. Hoods are not intended primarily for storage of chemicals. Materials stored in them should be kept to a minimum. Stored chemicals should not block vents or alter airflow patterns. Whenever practical, chemicals should be moved from hoods to vented cabinets for storage.

6. Solid objects and materials (such as paper) should not be permitted to enter the exhaust ducts of hoods as they can lodge in the ducts or fans and adversely affect their operation.

7. An emergency plan (see Section I.C.2) should always be prepared for the event of ventilation failure (power failure, for example) or other unexpected occurrence such as fire or explosion in the hood.

8. If laboratory workers are certain that adequate general laboratory ventilation will be maintained when the hoods are not running, hoods not in use should be turned off to conserve energy. If any doubt exists, however, or if toxic substances are being stored in the hood, the hood should be left on. Energy can also be conserved by the use of variable-volume hoods that modulate exhaust flow with sash position.

General Requirements for and Evaluation of Laboratory Hoods

The hood is the best-known local exhaust device used in laboratories. It is, however, but one part (typically the principal exhaust port) of the total ventilation system and should not be considered as separate from the total system, because its performance will be strongly influenced by other features in the general ventilation system. A number of recent studies of hood performance (see Section I.H.7) in terms of the protection hoods afford to laboratory workers have shown the importance of such factors as the volume of input air to the laboratory, the location of the laboratory input-air ports, the location of the hood within the laboratory, and the placement of apparatus within the hood. Any efforts to correct poor hood performance should involve consideration of all of these factors.

Hood Design and Construction

Laboratory hoods and the associated exhaust ducts should be constructed of nonflammable materials. They should be equipped with either vertical or horizontal sashes that can be closed. Welded steel construction is recommended for the sash frame. The glass within the sash should be laminated safety glass at least 7/32 in. thick or other equally safe material that will not shatter in the event of an explosion within the hood. The utility control valves, electrical receptacles (see Section I.G.1), and other fixtures should be located outside of the hood to minimize the need to reach within the hood proper. The construction materials, plumbing requirements, and interior design will vary, depending on the intended use of the hood. Although external baffles have been shown to provide improved directional airflow on some hoods, their use on all hoods is not indicated.

In recent years, the "supplementary-air hood" has become popular. This hood directs a blanket of untempered or partially tempered air vertical to the hood face between the operator and the sash. As much as 70% of the total air exhausted by the hood can be taken from a supplementary air source, resulting in considerable savings in energy. However, careful balancing of both the velocity and direction of the incoming air is required to achieve even air distribution across the hood face; if such a hood is operated with the sash closed, the supplementary air is of little value and may even upset the general laboratory ventilation. Consequently, the design and installation of a laboratory ventilation system employing supplementary-air hoods should not be attempted without the aid of a qualified ventilation engineer (see Section I.H.1).

Although hoods are most commonly considered as devices for controlling concentrations of toxic vapors, they can also serve to dilute and exhaust flammable vapors. Almost any hood can be used effectively or it can be overloaded or misused, resulting in spillage of vapor into the general room air. Also, an overloaded hood may contain an explosive mixture of air and a flammable vapor (see Section I.C.1). Both the hood designer and the user should be cognizant of this hazard and eliminate possible sources of ignition within the hood and its duct work if there is a potential for explosion. The concentration of vapors in the hood atmosphere can be estimated by using the procedures described above for calculating general room concentrations.

In some cases, the materials that might be exhausted by a hood are sufficiently toxic that they cannot be expelled into the air. Whenever possible, experiments involving such materials should be designed so that the toxic materials are collected in traps or scrubbers rather than being released into the hood. If, for some reason this is impossible, then HEPA filters are recommended for highly toxic particulates and activated charcoal filters can be used to adsorb highly toxic gases and vapors. Liquid scrubbers may also be used to remove both particulates and vapors and gases. None of these methods is completely effective. Incineration may be the ultimate method for destroying combustible compounds in exhaust air, but adequate temperature and dwell time are required to ensure complete combustion (see Section II.G.1). Incinerators require considerable energy, and other methods should be studied before resorting to their use. The optimum system for collecting or destroying toxic materials in exhaust air must be determined on a case-by-case basis. In all cases, such treatment of exhaust air should be considered only if it is not practical to pass the gases or vapors through a scrubber or adsorption train before they can enter the stream of hood exhaust air.

The ducts used for hood exhaust air should be dedicated to that purpose and not combined with other ventilation ducts within the building. It is usually best to have a separate duct for each hood to eliminate the possibility that toxic vapors exhausted into one hood could be channeled into an unused hood and reenter the general laboratory atmosphere. If several hoods are connected to a common exhaust duct, then some fail-safe arrangement should be provided to ensure that all of them are continually exhausting air when any one of them is in use.

Evaluation of Hood Performance

All hoods should be evaluated for performance when they are installed. New hood types or designs should be evaluated by a method [see, for example, the American Society of Heating, Refrigerating, and Air Conditioning Engineers *Handbook* or Chamberlin and Leahy (Section I.H.7)] that gives a quantitative rating of hood performance.

Performance should be evaluated against the design specifications for uniform airflow across the hood face as well as for the total exhaust air volume. Equally important is the evaluation of operator exposure or other purpose for prescribing the hood. This evaluation of hood performance should be repeated any time there is a change in any aspect of the ventilation system. Thus, changes in the total volume of input air, changes in the locations of air-input ports, or the addition of other auxiliary local ventilation devices [such as more hoods, vented cabinets, and snorkels (see Section I.H.3)] all call for reevaluation of the performance of all hoods in the laboratory.

The first step in the evaluation of hood performance is the use of a smoke tube or similar device to determine that the hood is on and exhausting air. The second step is to measure the velocity of the airflow at the face of the hood. The third step is to determine the uniformity of air delivery to the hood face by making a series of face velocity measurements taken in a grid pattern. Sets of measurements should be made with the hood sash fully opened and with the sash in one or more partially closed positions. The measuring instruments, anemometers or velometers (see Section I.H.6), should be calibrated before use. In measurement of airflow velocities at various points across the hood face, values for specific points may vary by ±25% from the average value. Greater variation than this should be corrected by adjustments in the interior hood baffle or, if necessary, by altering the path of the input air flowing into the room. Most laboratory hoods are equipped with a baffle that has movable slot openings at both the top and bottom. This baffle should be moved until the airflow (measured with an anemometer or velometer) approaches uniform. Larger hoods may require additional slots in the baffle to achieve uniform airflow across the hood face. If the sash is closed, the airflow will be less even.

The total volume of air being exhausted by a hood is the product of the average face velocity and the area of the hood face opening. If the hood and the general ventilating system are properly designed, face velocities in the range of 60-100 lfm will provide a laminar flow of air over the floor and sides of the hood. Higher face velocities (125 lfm or more), which exhaust the general laboratory air at a greater rate, are both wasteful of energy and likely to degrade hood performance by creating air turbulence at the hood face and within the hood. Such air turbulence can cause the vapors within the hood to spill out into the general laboratory atmosphere.

The rate of air exhaust from a hood is determined in part by the size and speed of the hood exhaust fan. In some cases, hoods are fitted with multiple-speed fan motors so that fan speed can be selected by using a switch on the hood face. Such controls, dampers or baffles, and other devices can be engineered into the hood to reduce the venting of conditioned air. However, in all cases, the designed air-exhaust rate of the hood will be achieved only if adequate input air is supplied to the laboratory. If the volume of input air is not sufficient, changes in the hood fan speed will do little to improve hood performance. Recently, in an effort to conserve energy, many laboratories have decreased the amount of input air, turned off some of the hoods, or done both during certain times of the day. Before such procedures are initiated, the effects on both hood performance and the overall laboratory ventilation system (see Section I.H.1) should be tested to avoid serious problems from inadequate ventilation. For example, if the hoods are the only exhaust ports in a laboratory, when they are all turned off, there may be no change of air in the laboratory. Alternatively, if all hoods are left on but the supply of input air is decreased, the performance of every hood in the building may be degraded to a potentially dangerous level because none of the laboratorics will have an adequate supply of input air. A hood that is not providing adequate ventilation performance is often worse than no hood at all because the laboratory worker is likely to have a false sense of security about the protection provided by the local ventilation system he or she is using. Assuming that the general ventilation system is properly designed, any decrease in the amount of input air being supplied should be accompanied by turning off selected hoods to maintain a balance between input and exhaust air.

The optimum face velocity of a hood (also called the capture velocity) will vary depending on its configuration. As noted above, too high a face velocity is likely to increase the turbulence within the hood and cause gases or vapors to spill from the hood into the room. It is necessary only that the capture velocity under use conditions be greater than cross currents of air at the hood face.

As the sash is closed, the hood face area is reduced and the capture velocity is increased. Thus, a hood that has a low airflow and a low capture velocity when the sash is open may provide excellent exposure control when the sash is closed. Likewise, a hood that has an adequate airflow but is operated with the sash open in a cross draft or other source of air turbulence may not provide adequate exposure control.

The second aspect of hood performance that should be evaluated is the presence or absence of air turbulence at the face of and within the hood. The observation of smoke patterns is used for this evaluation. Visible fumes or smoke can be generated by using cotton swabs dipped in titanium tetrachoride, commercial smoke tubes, or aerosol generators. Each hood should be tested for air turbulence and capture effectiveness before it is used and again after any change is made in the overall ventilation system of the laboratory. If there is excessive turbulence or if the hood fails to capture smoke, changes may be required in the hood face velocity, the location of the air-input ports, the physical location of the hood, or the volume of input air.

The location within the room of the hood will affect its performance. If the hood is placed so that cross drafts from the movement of people or air currents from open windows or doors exceed the capture velocity, material may be drawn from the hood into the room. Often, air turbulence at the hood face is best diminished by relocating the air-input ports or by adding external baffles near the hood face. A qualified ventilation engineer should be consulted for aid in solving such problems.

Another factor that influences the performance of a hood is the amount and location of equipment in it. As in the general laboratory, the air in the hood moves in all directions. Hot plates, heating mantles, and equipment standing in the hood may alter this movement and increase air turbulence. If the emission source is placed near the hood face, the vapors are likely to spill outside the hood. Although a high capture or face velocity may help prevent this spillage, a more satisfactory (and less energy-consuming) approach is to place the equipment farther into the hood. All equipment should be placed at least 10 cm (4 in.) back from the hood sash; in general, all equipment should be placed as far to the back of the hood as practical. In some laboratories, a colored stripe is painted on the hood work surface 10 cm back from the face to serve as a constant reminder. The less apparatus and bottles the hood contains the more likely that it will have laminar airflow across its working surface. Observing these simple precautions will often result in a significant improvement in hood performance at face velocities of 60-80 lfm. Because a substantial amount of energy is required to supply tempered input air to even a small hood, the use of hoods for the storage of bottles of toxic or corrosive chemicals is a very wasteful practice (see Sections II.B.4 and II.D.1 and 3), which can also, as noted above, seriously impair the effectiveness of the hood as a local ventilation device. Thus, it is preferable to provide separate vented cabinets for the storage of toxic or corrosive chemicals. The amount of air exhausted by such cabinets is much less than that exhausted by a properly operating hood.

The position and movement of the hood user will also affect the performance of the hood. A user standing in front of an open hood sash may cause considerable turbulence and eddy currents near the face of the hood. Placing equipment well back in the hood and partially closing the hood sash will help minimize losses caused by this air turbulence.

After the face velocity of each hood has been measured (and the air flow balanced if necessary), each hood should be fitted with an inexpensive manometer or other pressure-measuring device (or a velocity-measuring device) to enable the user to determine that the hood is operating as it was when evaluated. This pressure-measuring device should be capable of measuring pressure differences in the range of 0.1-2.0 in. of H_2O and should have the lower-pressure side connected to the duct just above the hood and the higher-pressure side open to the general laboratory atmosphere. Once such a device has been calibrated by measuring the hood face velocity, it will serve as a constant monitor of hood performance and can provide warning of inadequate hood performance that might arise from future defects or changes in the overall laboratory ventilation. In many cases, the indicator light on the front of a hood next to the switch simply indicates whether or not the switch is in the on position and provides no information about whether or not the hood fan motor is actually running or the input-air supply is adequate.

Perhaps the most meaningful (but also the most time-consuming) method of evaluating hood performance is to measure worker exposure while the hood is being used for its intended purpose. By using commercial personal air-sampling devices (see Section I.H.6) that can be worn by the hood user, worker exposure, both excursion peak and time-weighted average, can be measured using standard industrial hygiene techniques. The criterion for evaluating the hood should be the desired performance; i.e., does it contain vapors and gases at the desired worker-exposure level, and a sufficient number of measurements should be made to define a statistically significant maximum exposure based on worst-case operating conditions. Direct-reading instruments (see Section I.H.6) are available for determining the short-term concentration excursions that may occur in laboratory hood use. Various low-toxicity chemicals, such as Freon 11, sulfur hexafluoride, and certain detectable dyes, can also be used with direct-reading instrumentation to evaluate hood performance. However, for most purposes, the combination of a smoke test and a series of face velocity measurements will be sufficient to evaluate hood performance.

I.H.3 OTHER LOCAL EXHAUST SYSTEMS

The usual laboratory hood depends on the horizontal flow of a substantial volume of air to direct contaminated air away from the user. Although canopy hoods, snorkels, and similar devices that use vertical airflow are much less effective, they are useful in providing local ventilation above chromatographic and distillation equipment and various instruments (e.g., spectrometers) that cannot reasonably be placed in hoods. With good design, these vertical-airflow devices can be used to contain emissions of hazardous substances. Drop curtains or partial walls of plastic or metal to direct the airflow may make them more effective. However, without proper training, the users of these devices may find themselves in a contaminated air stream.

Whether the emission source is a vacuum-pump discharge vent, a gas chromatography exit port, or the top of a fractional distillation column, the local exhaust requirements are similar. The total airflow should be high enough to transport the volume of gases or vapors being emitted, and the capture velocity should be sufficient to collect the gases or vapors. The capture velocity is approximately 7.5% of the face velocity at a distance equal to the diameter of the local exhaust opening. Thus, a 3-in. snorkel or elephant trunk having a face velocity of 150 lfm will have a capture velocity of only 11 lfm at a distance of 3 in. from the opening. As the air-movement velocity in a typical room is 20 lfm, capture of vapors emitted 3 in. from the snorkel will be incomplete. However, vapors emitted at distances of 2 in. or less from the snorkel opening will be captured completely. Thus, these canopy-hood or snorkel devices will be effective only when the source of vapor emission is placed very close to the opening of the local exhaust system or if supplementary plastic or metal curtains are used to direct the airflow into the opening of the local exhaust device.

Despite these limitations, these canopy-hood and snorkel systems can provide useful and effective control of emissions of toxic vapors or dusts if they are installed and used correctly. It is not considered good practice to attach such devices to an existing hood duct; a separate exhaust duct should be provided. One very important consideration is the effect that such added local exhaust systems will have on the remainder of the laboratory ventilation (see Section I.H.1). Each snorkel or canopy hood added will be a new exhaust port in the laboratory and will compete with the existing exhaust ports for input air. Thus, before any extra local exhaust systems are added, a qualified ventilation engineer should be consulted. After such devices have been installed, all aspects of the ventilation system should be evaluated so that necessary changes in the volume of input air and in the location of input-air ports can be made. Failure to follow these precautions could result in serious degradation of all aspects of the ventilation system in the laboratory.

Downdraft ventilation has been used effectively to contain dusts and other dense particulates and high concentrations of heavy vapors that, because of their weight, tend to fall. Such systems require special engineering considerations to ensure that the particulates are transported in the air stream. Knock-outs or areas of low air velocity may be designed into such systems to permit collection of larger particles before the exhaust air is filtered or scrubbed.

I.H.4 SPECIAL VENTILATION AREAS

Glove Boxes and Isolation Rooms

Glove boxes are usually small units that have multiple ports in which arm-length rubber gloves are mounted, and the operator works through these. Construction materials vary widely, depending on the intended use. Clear plastic is frequently used because it allows visibility of the work area and is easily cleaned.

Glove boxes generally operate under negative pressure, so that any air leakage is into the box. If the material being used is sufficiently toxic to require the use of an isolation system, it is obvious that the exhaust air will require special treatment before release into the regular exhaust system. However, because these small units have a low airflow, scrubbing or adsorption (or both) can be accomplished with little difficulty. Some glove boxes operate under positive pressure. These boxes are commonly used for experiments in which protection from atmospheric moisture or oxygen is desired. If such glove boxes must be used with materials that present a high toxicity hazard, they should be thoroughly tested for leaks before each use and there should be a method of monitoring the integrity of the system (such as a shutoff valve or a pressure gage designed into it).

Isolation rooms use the same principles as glove boxes, except that the protected worker is within the unit. The unit itself operates under negative pressure, and the exhaust air requires special treatment before release. Many isolation units have a separate air supply to prevent cross-contamination. The workers enter the unit through clean rooms in which they remove their street clothes and don clean work clothes and other personal protective equipment, such as supplied-air respirators (see Section I.F.4). They reverse this procedure when leaving the isolation unit by removing their work clothes in a dirty room, passing through a shower, and then entering a clean room where they don street clothes.

These isolation areas require considerable engineering, and the training of personnel is most important. They are, however, frequently used for handling regulated carcinogens, although other equally potent but nonregulated carcinogens are being handled in regular laboratories by using carefully planned and monitored procedures and by following adequate safety procedures and local ventilation techniques (see Sections I.B.8 and 9). Isolation units should be used only for highly toxic substances that have physical properties that make control difficult or impossible by more conventional methods; other substances should be handled by general safety procedures.

Environmental Rooms

Environmental rooms (see also Section I.F.5), either as refrigeration cold rooms or as warm rooms for growth of organisms and cells, have the inherent property (as a result of their construction) of being a closed air-circulation system. Thus, the release of any toxic substance in these areas poses potential dangers. Also, because of the contained atmosphere in these rooms, there is a significant potential for the creation of aerosols and for cross-contamination of research projects. These problems should be controlled by preventing the release of aerosols or gases into the room environment.

Because of the contained atmosphere in environmental rooms, provision should be made to allow persons in them to escape rapidly. The doors of these rooms should be equipped with magnetic latches (preferable) or breakaway handles that would allow a trapped person to dismantle the door; the electrical system should be independent of the main power supply so that such persons are not confined in the dark.

As for other refrigerators, volatile flammable solvents should not be used in cold rooms (see Section I.G.4). The exposed motors for the circulating fans can serve as a source of ignition and initiate an explosion. The use of volatile acids should also be avoided in these rooms because such acids can corrode the cooling coils in the refrigeration system, which can lead to the development of leaks of toxic refrigerants.

I.H.5 MAINTENANCE OF VENTILATION SYSTEMS

Even the best engineered and installed ventilation system requires routine maintenance. Blocked or plugged air intakes and exhausts may alter the performance of the total ventilation system. Belts loosen, bearings require lubrication, motors need attention, ducts corrode, and minor components fail; these malfunctions, individually or collectively, can affect overall ventilation performance.

All ventilation systems should have a monitoring device that readily permits the user to determine whether the total system and its essential components are functioning properly. Manometers, pressure gages, and other devices that measure the static pressure in the air ducts are sometimes used to reduce the need for manual measuring of the airflow. The need for and the type of monitoring devices should be determined on a case-by-case basis. If the substance being contained has excellent warning properties and the consequence of overexposure is minimal, the system will need less stringent control than if the substance is highly toxic or has poor warning properties. The need for scheduled maintenance will also be determined by these factors.

I.H.6 SOURCES OF EQUIPMENT AND SUPPLIES FOR MEASURING LABORATORY VENTILATION SYSTEMS

Suppliers of Industrial Hygiene Sampling Equipment

1. MSA Research Corp., 600 Penn Center Boulevard, Pittsburgh, PA 15235
2. Applied Technology Division, E.I. duPont de Nemours & Co., Wilmington, DE 19898
3. Abcor Development Corp., 850 Main Street, Wilmington, MA 01887
4. Anatole J. Sipin Co., Inc., 425 Park Avenue, South New York, NJ 10016
5. Occupational Health and Safety Products Division, 3M Corp., Box 8327, Department BB1, St. Paul, MN 55113
6. Environmental and Process Instruments Division, Bendix Corp., Drawer 831, Lewisburg, WV 24901

Suppliers of Industrial Hygiene and Ventilation Monitoring Equipment

Velometers and Anemometers

1. Alnor Instrument Co., 7301 N. Caldwell Avenue, Niles, IL 60648
2. Davis Instrument Manufacturing Company, Inc., 513 East 36th Street, Baltimore, MD 21218
3. TSI Inc., Box 3394, St. Paul, MN 55164
4. Kurz Instruments, Inc., Box 849, Carmel Valley, CA 93924

Manometers, Gages, and Pitot Tubes

Dwyer Instruments, Inc., Box 373, Michigan City, IN 46360

Direct Reading Instruments

1. Foxboro Co., Wilkes Infrared Center, Box 449, South Norwalk, CT 06856
2. MDA Scientific Inc., 808 Busse Highway, Parkridge, IL 60068
3. HNU Systems, 30 Ossipee Road, Newton, MA 02164
4. Century Systems Corp., Box 818, Arkansas City, KA 67005
5. Gas-Tech, Inc., 331 Fairchild Drive, Mountain View, CA 94043

Indicators and Smoke Tubes

1. MSA Research Corp., 600 Penn Center Boulevard, Pittsburgh, PA 15235
2. Matheson Division, Will Ross, Inc., 1275 Valley Brook Avenue, Lyndhurst, NJ 07071
3. National Draeger, Inc., 401 Parkway View Drive, Pittsburgh, PA 15205
4. Environmental and Process Instruments Division, Bendix Corp., Drawer 831, Lewisburg, WV 24901

Aerosol Generators

1. Sierra Instruments, Inc., Drawer 909, Carmel Valley, CA 93924
2. Climet Instruments, 1320 West Colton Avenue, Redlands, CA 92373

I.H.7 SELECTED REFERENCES TO STUDIES OF LABORATORY HOODS FOR LOCAL EXHAUST VENTILATION

1. American Society of Heating, Refrigerating, and Air Conditioning Engineers. *Applications Handbook*, 1978, Chapter 15.
2. Bolton, N. E.; Porter, W. E.; Alcorn; S. P.; Everett, W. S.; Hunt, J. B. *Minimum Acceptable Face Velocities of Laboratory Fume Hoods and Guidelines for Their Classification*; Oak Ridge National Laboratory, Oak Ridge, TN, 1978, Report ORNL/TM-6400.
3. Caplan, K. J.; Knutson, G. W. *The Effect of Room Air Challenge on the Efficiency of Laboratory Fume Hoods*; ASHRAE Trans. 83, 1977, Part 1.
4. Chamberlin, R. I., Leahy, J. E. Laboratory Fume Hood Standards; Facilities Engineering and Real Properties Branch, USEPA, 1978, Contract No. 68-01-4661.
5. Clayton, George D.; Clayton, Florence, E., Eds. *Patty's Industrial Hygiene and Toxicology*, 3rd ed.; John Wiley and Sons: New York, 1978; Vol. 1.
6. Committee on Industrial Ventilation. *Industrial Ventilation*; 15th ed.; American Conference of Government Industrial Hygienists: Cincinnati, OH, 1978.
7. Fuller, F. H.; Etchells, A. W. *Safe Operation with the 0.3 m/s (60 fpm) Laboratory Hood*; Am. Soc. of Heating, Refrigerating, and Air Conditioning Engineers Journal, Oct. 1979, 49.
8. Hemen, W. C. L. *Plant and Process Ventilation*, 2nd ed.; Industrial Press: New York, 1955 (revised 1963).
9. Olishifski, Julian B., Ed. *Fundamentals of Industrial Hygiene*, 2nd ed.; National Safety Council: Chicago, 1979.
10. Peterson, J. E. *An Approach to a Rational Method of Recommending Face Velocities for Laboratory Hoods*; Am. Ind. Hyg. Asso. J., 1959, 20, 259.

11. Peterson, J. E.; Peay, J. A. *Laboratory Fume Hoods and Their Exhaust Systems*; Air Conditioning, Heating, and Ventilation, May 1963, 60, 63.

12. TLV Airborne Contaminants Committee. *Threshold Limit Values for Chemical Substances and Physical Agents in the Workroom Environment with Intended Changes for 1979*; American Conference of Governmental Industrial Hygienists: Cincinnati, OH, 1979.

(See also Sections I.E.3 and I.A.18).

II

PROCEDURES FOR THE PROCUREMENT, STORAGE, DISTRIBUTION, AND DISPOSAL OF CHEMICALS

II.A

Procedures for Ordering and Procurement of Chemicals

The achievement of safe handling, use, and disposal of hazardous substances begins with the persons who requisition such substances and those who approve their purchase orders. These persons must be aware of the potential hazards of the substances being ordered, know whether or not adequate facilities and trained personnel are available to handle such substances, and should ensure that a safe disposal route exists.

Before a new substance that is known or suspected to be hazardous is received, information concerning its proper handling methods, including proper disposal procedures, should be given to all those who will be involved with it. If the distribution system involves receiving room or storeroom personnel, they should be advised that the substance has been ordered. It is the responsibility of the laboratory supervisor to ensure that the facilities are adequate and that those who will handle any material have received proper training and education to do so safely.

For a large number of substances, Material Safety Data Sheets (OSHA form 20), which give physical property data and toxicological information, can be obtained by request to the vendor. However, the quality and depth of information on these sheets varies widely. For substances for which such sheets are not available, e.g., research chemicals usually sold in small quantities, the manufacturer, if contacted directly, will usually provide whatever health and safety information is available.

The U.S. Department of Transportation (DOT) requires that shippers furnish and attach department-prescribed labels (see color insert) on all shipments of hazardous substances. These labels indicate the nature of the hazard(s) of the substance(s) shipped and thus provide some indication to receiving room, storeroom, and stockroom personnel of the type of hazard received, but should not be relied on after the container has been opened.

Because storage in laboratories is usually restricted to small containers (see Section II.D.2), it is sometimes preferable to order in small-container lots to avoid the hazards associated with repackaging. Some chemical suppliers ship solvents in small metal containers to avoid the hazard of breakage.

It is preferable that all substances be received at a central location for distribution to the storerooms, stockrooms, and laboratories. Central receiving is also helpful in monitoring substances that may eventually enter the waste disposal system. An inventory of substances kept in the storerooms and stockrooms can serve to alert those responsible for disposal as to what they may expect regarding the quantity and nature of substances they may be required to handle.

No container of a chemical or cylinder of a compressed gas should be accepted that does not have an identifying label. For chemicals, it is desirable that this label correspond to ANSI Z129.1, which requires, at a minimum, the following components:

1. identification of contents of container;
2. signal word and summary description of any hazard(s);
3. precautionary information—what to do to minimize hazard or prevent an accident from happening;
4. first aid in case of exposure;
5. spill and cleanup procedures; and
6. if appropriate, special instructions to physicians.

Receiving room, storeroom, and stockroom personnel should be knowledgeable about or trained in the handling of hazardous substances. Such training should include the physical handling of containers of chemicals so that they are not dropped, bumped, or subject to crushing by being piled one upon another. Information should be provided about environmental and hazard-initiating exposures that must be avoided. Some of the more common items with which receiving room, storeroom, and stockroom personnel need to be familiar include the following:

1. the use of proper material handling equipment, protective apparel (see Sections I.F.1-3), and safety equipment (see Section I.F.4);
2. emergency procedures, including the cleanup of spills (see Sections II.E.5 and I.F.5) and the disposal of broken containers;

3. the dangers of contacting chemicals by skin absorption, inhalation, or ingestion (see Section I.B.1);

4. the meanings of the various DOT labels on shipping packages;

5. the proper methods of material handling and storage, especially the incompatibility of some common substances, the dangers associated with alphabetical storage, and the sensitivity of some substances to heat, moisture, light, and other storage hazards (see Table 2 and Section I.C.3);

6. the special requirements of heat-sensitive materials, including those shipped refrigerated or packed in dry ice;

7. the problems associated with compressed gases, including unique situations such as the construction of an acetylene cylinder (see Sections II.B.6 and I.D.1);

8. the hazards associated with flammable liquids (especially the danger of their vapors catching fire some distance from the container) and explosives (see Chapter I.C) and of toxic gases and vapors and oxygen displacement (see Chapter I.B);

9. substances that react with water, giving rise to hazardous conditions (e.g., alkali metals, burning magnesium, metal hydrides, acid chlorides, phosphides, and carbides) (see Section I.C.3);

10. the federal and state regulations governing controlled substances such as radioactive materials, drugs, ethyl alcohol, explosives, and needles and syringes;

11. chemicals that have offensive smells; and

12. packages that exhibit evidence that the inside container has broken and leaked its contents.

II.B

Procedures for Storing Chemicals in Storerooms and Stockrooms

There is a range of possibilities for storing chemical substances. The arrangements made will depend on the size of the organization, the quantities handled, and the nature of the problems.

Often, the provision of adequate storage space is given little consideration in the design of laboratory buildings. Lack of sufficient storage space can create hazards due to overcrowding, storage of incompatible chemicals together, and poor housekeeping. Adequate, properly designed and ventilated storage facilities should be provided to ensure personnel safety and property protection.

In many instances, chemicals are delivered after receipt in the institution directly to the individual who initiated the order. If the facilities of the laboratory are appropriate (see Chapter II.D) for the kinds and quantities of materials used, this system may be eminently satisfactory.

However, experience has shown that it is usually necessary to maintain a reserve of supplies in excess of the amounts that can be kept safely in the laboratory. If the quantities are large or the volumes of the individual containers are such that repackaging is necessary, then a safe place is needed to store these containers and to perform these functions. Depending on needs, this could be a stockroom for the laboratory or a central storeroom for the particular organization.

Stored chemicals should be examined at periodic intervals (at least annually). At this time, those that have been kept beyond their appropriate shelf life or have deteriorated, have questionable labels, are leaking, have corroded caps, or have developed any other problem should be disposed of in a safe manner (see Chapters II.E and G). A first-in, first-out system of stock keeping should be used.

Shelved chemicals can walk, creep, and even tip over. Such chemicals can be prevented from falling off by placing retaining shock cords or similar restraining devices across the open face of the shelf or by raising the forward face of the shelf about one-quarter inch.

II.B.1 STOCKROOM DESIGN

Stockrooms are similar to central storerooms except that the quantities of materials involved are usually much smaller (the materials stored will be the inventory for a particular laboratory or groups of laboratories) and such rooms are usually within or close to the areas served.

Stockrooms should not be used as preparation areas because of the possibility that an accident will occur and thereby unnecessarily contaminate a large quantity of materials. Preparation and repackaging should be performed in a separate area.

Stockrooms should be conveniently located and open during normal working hours so that laboratory workers need not store excessive quantities of chemicals in their laboratories. However, this does not imply that all laboratory workers should have unlimited access to the chemicals in the stockroom. Procedures must be established for the operation of any stockroom that place responsibility for its safety and inventory control in the hands of one person. If it is not feasible to have a full-time stockroom clerk, then one person who is readily available should be assigned that responsibility.

Stockrooms should be well ventilated. If storage of opened containers is permitted, extra local exhaust ventilation and the use of outside storage containers or spill trays are necessary.

II.B.2 FLAMMABLE LIQUIDS

Centralized storage of bulk quantities of flammable liquids provides the best method of controlling the associated fire hazard.

Because the most effective way to minimize the impact of a hazard is to isolate it, a storage and dispensing room for flammable liquids is best located in a special building separated from the main building. If this is not feasible, and the room must be located in a main building, the preferred location is a cutoff area on the at-grade level and having at least one exterior wall. [NOTE: Cutoff is a fire-protection term defined as "separated from other areas by fire-rated construction."] In any case, storage rooms for flammable liquids should not be placed on the roof, located on a below-grade level, an upper floor, or in the center of the building. All of these locations are undesirable because they are less accessible for fire fighting and potentially dangerous to the safety of the personnel in the building.

The walls, ceilings, and floors of an inside storage room for flammable liquids should be constructed of materials having at least a 2-hour fire resistance, and there should be self-closing Class B fire doors [see the OSHA standard (29 CFR 1910.106(d)(4)(i)) or the National Fire Protection Association (NFPA) Standard (No. 30.4310)]. All storage rooms should have adequate mechanical ventilation controlled by a switch outside the door and explosion-proof lighting and switches. Other potential sources of ignition, such as burning tobacco and lighted matches, should be forbidden.

II.B.3 DRUM STORAGE

Fifty-five gallon drums are commonly used to ship flammable liquids but are not intended as long-time inside storage containers. It is not safe to dispense from sealed drums exactly as they are received. The bung should be removed and replaced by an approved pressure and vacuum relief vent to protect against internal pressure buildup in the event of fire or if the drum might be exposed to direct sunlight.

If possible, drums should be stored on metal racks placed such that the end bung openings are toward an aisle and the side bung openings are on top. The drums, as well as the racks, should be grounded with a minimum length of American wire gage 10 wire. Because effective grounding requires metal-to-metal contact, all dirt, paint, and corrosion must be removed from the contact areas. Spring-type battery clamps and a minimally sized conductor (e.g., American wire gage 8 or 10) are satisfactory. It is also necessary to provide bonding to metal receiving containers to prevent accumulation of static electricity (which will discharge to the ground, creating a spark that could ignite the flammable vapors). Drip pans that have flame arresters should be installed or placed under faucets.

Dispensing from drums is usually done by one of two methods. The first is gravity based through drum faucets that are self-closing and require constant hand pressure for operation. Faucets of plastic construction are not generally acceptable due to chemical action on the plastic materials.

The second, and safer, method is to use an approved hand-operated rotary transfer pump. Such pumps have metering options and permit immediate cutoff control to prevent overflow and spillage, can be reversed to siphon off excess liquid in case of overfilling, and can be equipped with drip returns so that any excess liquid can be returned to the drum.

II.B.4 TOXIC SUBSTANCES

Toxic substances should be segregated from other substances and stored in a well-defined or identified area that is cool, well ventilated, and away from light, heat, acids, oxidizing agents, moisture, and such.

The storage of unopened containers of toxic substances normally presents no unusual requirements. However, because containers occasionally develop leaks or are broken, storerooms should be equipped with exhaust hoods or equivalent local ventilation devices (see Section I.H.4) in which containers of toxic substances can be handled.

Opened containers of toxic substances should be closed with tape or other sealant before being returned to the storeroom and should not be returned unless some type of local exhaust ventilation is available. (See Section I.B.8 for specific storage and record-keeping procedures for compounds of known high chronic toxicity.)

II.B.5 WATER-SENSITIVE CHEMICALS

Some chemicals react with water to evolve heat and flammable or explosive gases. For example, potassium and sodium metals and many metal hydrides react on contact with water to produce hydrogen, and these reactions evolve sufficient heat to ignite the hydrogen with explosive violence. Certain polymerization catalysts, such as aluminum alkyls, react and burn violently on contact with water.

Storage facilities for water-sensitive chemicals should be constructed to prevent their accidental contact with water. This is best accomplished by eliminating all sources of water in the storage area; for example, areas where large quantities of water-sensitive chemicals are stored should not have automatic sprinkler systems. Storage facilities for such chemicals should be of fire-resistant construction, and other combustible materials should not be stored in the same area.

II.B.6 COMPRESSED GASES

Cylinders of compressed gases should be stored in well-ventilated, dry areas. Where practicable, storage rooms should be of fire-resistant construction and above ground. Cylinders may be stored out of doors, but some protection must be provided to prevent corrosion of the cylinder bottom and air circulation must not be restricted. All storage and use of compressed gases should be in compliance with OSHA regulations (29 CFR 1910.166-171).

Compressed gas cylinders should not be stored near sources of ignition nor where they might be exposed to corrosive chemicals or vapors. They should not be stored where heavy objects might strike or fall on them, such as near elevators, service corridors, and unprotected platform edges.

The cylinder storage area should be posted with the names of the gases stored. Where gases of different types are stored at the same location, the cylinders should be grouped by type of gas (e.g., flammable, toxic, or corrosive). If possible, however, flammable gases should be stored separately from other gases and provision should be made to protect them from fire. Full and empty cylinders should be stored in separate portions of the storage area, and the layout should be arranged so that older stock can be used first with minimum handling of other cylinders.

Cylinders and valves are usually equipped with various safety devices, including a fusible metal plug that melts at 70-95°C. Although most cylinders are designed for safe use up to a temperature of 50°C, they should not be placed where they can become overheated (e.g., near radiators, steam pipes, or boilers). Cylinder caps to protect the container withdrawal valve should be in place at all times during storage and movement to and from storage.

Cylinders should be stored in an upright position where they are unlikely to be knocked over, or they should be secured in an upright or horizontal position. Acetylene cylinders should always be stored valve end up to minimize the possibility of solvent discharge (see Section I.D.4). Oxygen should be stored in an area that is at least 20 ft away from any flammable or combustible materials (especially oil and grease) or separated from them by a noncombustible barrier at least 5 ft high and having a fire-resistance rating of at least 1/2 hour.

Cylinders are sometimes painted by the vendor to aid in the recognition of their contents and make separation of them during handling easier. However, this color coding is not a reliable method for identification of their contents; the stenciled or printed name on the cylinder is the only accepted method. If it is suspected that a stored cylinder is leaking, the procedures described in Section II.E.6 should be followed.

II.C
Procedures for Distributing Chemicals from Stockrooms to Laboratories

The method of transport of chemicals between stockrooms and laboratories must reflect the potential danger posed by the specific substance.

II.C.1 CHEMICALS

When chemicals are hand carried, they should be placed in an outside container or acid-carrying bucket to protect against breakage and spillage. When they are transported on a wheeled cart, the cart should be stable under the load and have wheels large enough to negotiate uneven surfaces (such as expansion joints and floor drain depressions) without tipping or stopping suddenly.

To avoid exposure to persons on passenger elevators, if possible, chemicals should be transported on freight-only elevators.

Provisions for the safe transport of small quantities of flammable liquids include (a) the use of rugged pressure-resistant, nonventing containers, (b) storage during transport in a well-ventilated vehicle, and (c) elimination of potential ignition sources.

II.C.2 CYLINDERS OF COMPRESSED GASES

The cylinders that contain compressed gases are primarily shipping containers and should not be subjected to rough handling or abuse. Such misuse can seriously weaken the cylinder and render it unfit for further use or transform it into a rocket having sufficient thrust to drive it through masonry walls. To protect the valve during transportation, the cover cap should be left screwed on hand tight until the cylinder is in place and ready for actual use. Cylinders should never be rolled or dragged. The preferred transport, even for short distances, is by suitable hand truck with the cylinder strapped in place. Only one cylinder should be handled at a time.

II.D

Procedures for Storing Chemicals in Laboratories

The amounts of toxic, flammable, unstable, or highly reactive materials that should be permitted in laboratories are an important concern. To arbitrarily restrict quantities may interfere with laboratory operations but, conversely, unrestricted quantities can result in the undesirable accumulation of such materials in the laboratory. It is, however, necessary to comply with any local statutory restrictions on allowable quantities.

It is necessary to balance the needs of the laboratory workers and the established requirements for safety. Decisions in this area will be affected by the level of competence of the workers, the level of safety features designed into the facility, the location of the laboratory, the nature of the chemical operations, and the accessibility of the stockroom. In some cases, local regulations or insurance requirements will determine the quantities that can be stored. In general, all laboratories should have two exits (one may be an emergency exit) so that a fire at one exit will not block occupants' escape; doors that open outward are desirable.

II.D.1 GENERAL CONSIDERATIONS

Every chemical in the laboratory should have a definite storage place and should be returned to that location after each use.

The storage of chemicals on bench tops is undesirable; in such locations, they are unprotected from potential exposure to fire and are also more readily knocked over. Storage in hoods is also inadvisable because this practice interferes with the air flow in the hood (see Sections I.H.3 and 4), clutters up the working space, and increases the amount of materials that could become involved in a hood fire.

Storage trays or secondary containers should be used to minimize the distribution of material should a container break or leak.

Because most laboratory workers tend to store hazardous materials in the cabinet space under the hood, the provision of ventilated cabinets in this location is advisable. The use of such cabinets also has the advantage that, because of proximity to the hood, the safe practice of making transfers of hazardous materials in the hood is encouraged.

As for storerooms and stockrooms, care should be taken in the laboratory to avoid exposure of chemicals to heat or direct sunlight and to observe precautions regarding the proximity of incompatible substances (see Chapter II.A, Section I.C.3, and Table 2)

Laboratory refrigerators are to be used for the storage of chemicals only; food must not be placed in them. All containers placed in the refrigerator should be properly labeled (identification of contents and owner, date of acquisition or preparation, and nature of any potential hazard) and, if necessary, should be sealed to prevent escape of any corrosive vapors. Flammable liquids should not be stored in laboratory refrigerators unless the unit is an approved, explosion-proof, or laboratory-safe type (see Section I.G.4 and NFPA Standards 45 and 56D).

The chemicals stored in the laboratory should be inventoried periodically, and unneeded items should be returned to the stockroom or storeroom. At the same time, containers that have illegible labels and chemicals that appear to have deteriorated should be disposed of (see Chapters II.E and G).

On termination, transfer, graduation, or such of any laboratory personnel, those personnel and the laboratory supervisor should arrange for the removal or safe storage of all hazardous materials those persons have on hand.

II.D.2 FLAMMABLE LIQUIDS

OSHA regulations for the laboratory storage of flammable and combustible liquids are not based on fire prevention and protection principles but rather address the types and sizes of containers allowable [and would permit the storage of 60-gal metal drums in laboratories of colleges and universities (see Section II.B.3)]. The NFPA standard (No. 45), on the other hand, has a quantity limit per 100 ft^2 that depends on the construction and fire protection afforded in the laboratory and restricts instructional laboratories to half the quantities for industrial or graduate student laboratories. A second NFPA standard (No. 30) addresses the amounts that may be stored outside of an approved flammable-liquid storage room or cabinet, but does not consider fire protection features available (see Table 7).

Whenever feasible, quantities of flammable liquids greater than 1 liter should be stored in metal containers. Portable approved safety cans are one of the safest methods of storing flammable liquids. These cans are available in a variety of sizes and materials. They have spring-loaded spout covers that can open to relieve internal pressure when subjected to a fire and will prevent leakage if tipped over. Some are equipped with a flame arrester in the spout that will prevent flame propagation into the can. If possible, flammable liquids received in large containers should be repackaged into safety cans for distribution to laboratories (see Section II.B.2). Such cans must be properly labeled to identify their contents.

Small quantities of flammable liquids should be stored in ventilated storage cabinets made of 18-gage steel and having riveted and spot-welded seams. Such cabinets are of double-wall construction and have a 1.5-in. air space between the inner and the outer walls. The door is 2 in. above the bottom of the cabinet, and the cabinet is liquid tight to this point. It is provided with vapor-venting provisions and can be equipped with a sprinkler system. (Materials that react with water should not be stored in sprinkler-equipped cabinets.) Some models have doors that close automatically in the event of fire.

If, for reasons of cost or space limitations, storage cabinets must be constructed of wood, they should be built according to the Los Angeles Fire Department specifications (see also NFPA Standard 30).

In any case, the hazard of storage of flammable materials in wooden cabinets in existing laboratories can be decreased by the use of intumescent fire-retardant coatings or other means that provide effective fire insulation. [NOTE: On heating, intumescent materials expand from a thin paintlike coating to a thick puffy coating that insulates or excludes oxygen and protects the subsurface from ignition.]

Other considerations in the storage of flammable liquids in the laboratory include ensuring that aisles and exits are not blocked in the event of fire; that accidental contact with strong oxidizing agents such as chromic acid, permanganates, chlorates, perchlorates, and peroxides is not possible; and that sources of ignition are excluded.

II.D.3 TOXIC SUBSTANCES

Chemicals known to be highly toxic, including those classified as carcinogens, should be stored in ventilated storage areas in unbreakable chemically resistant secondary containers.

TABLE 7 Container Size Limitations for Flammable and Combustible Liquids

	Flammable Liquids						Combustible Liquids			
	Class IA		Class IB		Class IC		Class II		Class IIIA	
Type of Container	Liters	Gal	Liters	Gal	Liters	Gal	Liters	Gal	Liters	Gal
Glass	0.5	0.12	1	0.25	4	1	4	1	4[a]	1[a]
Metal (other than DOT drums)	4	1	20	5	20	5	20	5	20	5
Safety cans	7.5	2	20	5	20	5	20	5	20	5
Metal drums (DOT specifications[b])	225	60	225	60	225	60	225	60	225	60
Approved portable tanks[c]	2500	660	2500	660	2500	660	2500	660	2500	660

[a]OSHA limitation; NFPA Nos. 30 and 45 allow 20 liters (5 gal).

[b]Maximum size permitted in a laboratory for Class I materials is 20 liters (5 gal); drum size is permitted only in an inside storage room (OSHA 1910.106 and NFPA No. 30).

[c]Permitted only outside of buildings.

Only minimum working quantities of toxic materials should be present in the work area. Storage vessels containing such substances should carry a label such as the following: CAUTION: HIGH CHRONIC TOXICITY or CANCER-SUSPECT AGENT (see Sections I.B.8 and 9).

Storage areas for substances that have high acute or chronic toxicity should exhibit a sign warning of the hazard, have limited access, and be adequately ventilated (see Sections I.B.7 and 5). An inventory of these toxic materials should be maintained. For those chronically toxic materials designated as regulated carcinogens, this inventory is required by federal and state regulations. Adequate ventilation is of particular concern for hazardous materials that have a high vapor pressure (such as bromine, mercury, and mercaptans).

II.D.4 COMPRESSED GASES

Cylinders of compressed gases should be securely strapped or chained to a wall or bench top to prevent their being knocked over accidentally. When they are not in use, it is good practice to keep them capped. Care should be taken to keep them away from sources of heat or ignition (see also Sections II.B.6 and I.D.1).

II.D.5. SELECTED BIBLIOGRAPHY

1. Compressed Gas Association. *Safe Handling of Compressed Gases in Containers* and *Characteristics and Safe Handling of Medical Gases*; New York, 1965, Pamphlets P1 and P2.

2. Gatson, P. J. *Care, Handling, and Disposal of Dangerous Chemicals*; Northern Publishers: Aberdeen, Scotland, 1970.

3. Henry, R. J.; Olitzky, I.; Lee, N. D.; Walker, B.; Beattie, J. *Safety in the Clinical Laboratory*, 1st ed.; Bio-Science Enterprises: Van Nuys, Calif.

4. Los Angeles Fire Department. *Hazardous Materials Storage Cabinets*; Los Angeles, Calif., Jan. 1, 1960, Standard 40.

5. Manufacturing Chemists Association. *Guide for Safety in the Chemical Laboratory*; Van Nostrand-Reinhold: New York.

6. Matheson Gas Products. *Safe Handling of Compressed Gases in Laboratory and Plant*; East Rutherford, N.J.

7. National Fire Protection Association. No. 30: *Flammable and Combustible Liquids Code*; No. 325M: *Fire Hazard Properties of Flammable Liquids, Gases, and Volatile Solids*; No. 45: *Fire Protection for Laboratories Using Chemicals*; No. 49: *Hazardous Chemical Data*; No. 491M: *Manual of Hazardous Chemical Reactions*; and SPR-51: *Flash Point Index of Trade Name Liquids*; Boston, Mass.

II.E
Procedures for Disposing of Chemicals in Laboratories

Proper disposal of the substances they have used is an important responsibility of all laboratory workers. Arrangements for disposal may vary from laboratory to laboratory, depending on the facilities and the types of substances used, but the basic principle is that substances must be disposed of in ways that avoid harm to people and the environment. Wastes should be transferred in a form that is safe and acceptable to the people involved in disposal operations. It is also important to consider the future fate of the waste substance. (See Chapter II.G.)

II.E.1 GENERAL CONSIDERATIONS

The plan for safe disposal of the substances used is as much a part of the plan for the experiment as is the acquisition of materials, the experimental procedures, and the isolation or storage. If an experiment involves new types of disposal problems, the laboratory worker should discuss the disposal plan with the laboratory supervisor and, if necessary, with the safety coordinator for the organization.

If practical, very hazardous substances should be converted to less hazardous substances in the laboratory rather than being placed directly in containers. For example, strong carcinogens should be oxidized in solution in the laboratory before disposal, and highly reactive substances, such as metallic sodium and peroxides, should be converted to less reactive substances. Reactions may be moderated by dilution, cooling, or the slow addition of a neutralizing agent. For water-miscible materials, pouring the reaction mixture onto a bed of ice can often be a way to cool and dilute it simultaneously.

All persons using chemicals in the laboratory should be generally aware of the toxic properties of the substance(s) used, including consideration of the toxic properties of possible reaction products (see Chapter I.E.1). If the toxic properties of possible products are not known, the products should be treated with respect and the disposal method should take account of the uncertain hazards. Some products may be disposed of as an integral part of the experiment (e.g., by using a scrubber for a gaseous product).

The cleanup of a laboratory that has been damaged by fire may present special hazards, e.g., the presence of toxic chemicals in the atmosphere and work environment, and may require special precautions.

The disposal of chemicals from instructional laboratories is a special problem because the students in such laboratories are inexperienced, the quantities of wastes may be relatively large, and the facilities may not be optimum. Education of students about how to handle chemicals and dispose of them safely is an integral part of their laboratory training.

II.E.2 DISPOSAL TO THE SEWER SYSTEM

Sewer systems operate in various ways, and some of them may be harmed or may present hazards for people and the environment when some chemicals are added directly to them. Generally, there are local regulations about what may be poured down the drain. The laboratory supervisor should know the local regulations and communicate this information to the laboratory workers so that they can conform to the regulations. All laboratory workers should know and respect these regulations. For sewer systems that discharge into waterways, federal regulations limit the disposal of certain toxic chemicals. In general, the following rules regarding disposal into a sewer system should be followed:

1. Only water-soluble substances should be disposed of in the laboratory sink. Solutions of flammable solvents must be sufficiently dilute that they do not pose a fire hazard.
2. Strong acids and bases should be diluted to the pH 3-11 range before they are poured in the sewer system. Acids and alkalis should not be poured into the sewer drain at a rate exceeding the equivalent of 50 ml of concentrated substance per min.
3. Highly toxic, malodorous, or lachrymatory chemicals should not be disposed of down the drain. Laboratory drains are generally interconnected; a substance that goes down one sink may well come up as a vapor in another. Sinks are usually communal property, and there is a very real

hazard of chemicals from two sources contacting one another; the sulfide poured into one drain may contact the acid poured into another, with unpleasant consequences for all in the building. Some simple reactions can even cause explosions (e.g., ammonia plus iodine, silver nitrate plus ethanol, or picric acid plus lead salts).

4. Small amounts of some heavy-metal compounds may be disposed of in the sink, but larger amounts may pose a hazard for the sewer system or water supply.

(See Chapter II.G.)

II.E.3 DISPOSAL OF SOLID CHEMICAL WASTES

Each organization should have procedures for collecting solid chemical wastes from the laboratories and arranging for disposal by the institution (see Chapter II.G). These procedures should include a clear understanding as to who is in charge and what the responsibilities of the laboratory workers are with respect to the identification of hazards that may be encountered in handling, transporting, and disposing of the solid waste. The people picking up such material should be aware of the hazards and know what to do in case of a spill during transportation.

The solid chemical wastes of a laboratory should be placed in containers provided for that purpose. When bottles are used, they should be placed in buckets. It is always important to be sure that all wastes are adequately labeled. The laboratory worker should be aware of the hazards that may be involved in disposing of particular solid chemical wastes and the importance of segregating incompatible materials (see Section I.C.3 and Table 2).

II.E.4 DISPOSAL OF LIQUID CHEMICAL WASTES

Similar to the requirements for solid chemical wastes, each organization should have a procedure for collecting liquid chemical wastes from the laboratories and arranging for their disposal by the institution (see Chapter II.G). Suitable containers should be provided, and the laboratory workers should understand what may, or may not, be placed in these containers and which materials require special labeling.

Waste solvents that are free of solids and corrosive or reactive substances may be collected in a common bottle or can, which is taken away when full. If this system is used, it is essential to consider exactly what mixtures will go into the can and whether the substances involved are compatible (this may include waste from the neighboring laboratory should its waste solvent bottle be full). Segregation into two or three types of waste is often useful (e.g., chlorinated solvents, hydrocarbons), as is the use of completely separate bottles for waste that poses special difficulties. In particular, because chlorinated solvents form hydrogen chloride on combustion, they often must be segregated from materials destined for incineration as their burning will violate local air pollution ordinances. Generally speaking, separated and well-defined waste is easier to dispose of and, if an outside contractor is used, is also less expensive. All wastes posing hazards should be so labeled.

Some solvents (such as ethers and secondary alcohols) form explosive peroxides on standing. Some reactions can cause explosions directly (e.g., acetone plus chloroform in the presence of a base). Others, such as acid-base interactions, can generate sufficient heat to vaporize or ignite flammable materials such as carbon disulfide. The addition of hot materials can cause the buildup of pressure in a tightly closed solvent container, with the potential for compressive ignition. The acid formed when halogenated solvents are left moist can corrode cans, as can any dissolved corrosive in a discarded mixture.

When large quantities of a solvent are involved, consideration should be given to recycling rather than disposal. This operation also involves some potential hazard and expense, but these limitations may be less severe than those for disposal, especially as disposal costs are increasing.

II.E.5 DISPOSAL OF ESPECIALLY HAZARDOUS WASTES

This class of chemical wastes includes very toxic substances, strong carcinogens, mutagens, nerve gases, explosives, and substances in tanks and other sealed containers. The laboratory worker has the responsibility to ensure that proper arrangements for disposal of these materials are made. Wherever possible, chemical reaction in the laboratory to produce less hazardous substances should be undertaken. For the case of those chemicals regulated as carcinogens, EPA disposal rules must be followed (see Chapter II.G).

A spill of one of these substances can be an especially serious hazard. Personnel working with such substances should have contingency plans, equipment, and materials available for coping with potential accidents.

II.E.6 SPILLS

Experience has shown that the accidental release of hazardous substances is a common enough occurrence to require preplanning for procedures that will minimize exposure of personnel and property. Such procedures may range from having available a sponge mop and bucket to having an emergency spill-response team, complete with protective apparel (see Sections I.F.1-3), safety equipment (see Section I.F.4), and materials to contain, confine, dissipate, and clean up the spill.

The preplanning should include consideration of the following factors:

1. potential location of the release (e.g., outdoors versus indoors; in a laboratory, corridor, or storage area, on a table, in a hood, or on the floor),
2. the quantities of material that might be released and whether the substance is a piped material or a compressed gas,
3. chemical and physical properties of the material (e.g., its physical state, vapor pressure, and air or water reactivity),
4. hazardous properties of the material (its toxicity, corrosivity, and flammability), and
5. the types of personal protective equipment that might be needed.

In any event, there should be supplies and equipment on hand to deal with the spill, consistent with the hazards and quantities of the spilled substance. These cleanup supplies should include neutralizing agents (such as sodium carbonate and sodium bisulfate) and absorbants (such as vermiculite and sand). Paper towels and sponges may also be used as absorbent-type cleanup aids, although this should be done cautiously. For example, paper towels used to clean up a spilled oxidizer may later ignite, and appropriate gloves (see Section I.F.2) should be worn when wiping up highly toxic materials with paper towels. Also, when a spilled flammable solvent is absorbed in vermiculite or sand, the resultant solid is highly flammable and gives off flammable vapors and, thus, must be properly contained or removed to a safe place.

Commercial spill kits are available that have instructions, absorbents (one of these, SOLUSORB®, not only absorbs spilled liquid but also reduces its vapor pressure to a relatively safe level and thus reduces the fire hazard), reactants, and protective equipment. These kits may be located strategically around work areas much as fire extinguishers are.

If a spill does occur, the following general procedures may be used but should be tailored to individual needs:

1. Attend to any persons who may have been contaminated.
2. Notify persons in the immediate area about the spill.
3. Evacuate all nonessential personnel from the spill area.
4. If the spilled material is flammable, turn off ignition and heat sources.

5. Avoid breathing vapors of the spilled material; if necessary, use a respirator (see Section I.F.4).
6. Leave on or establish exhaust ventilation if it is safe to do so (see Chapter I.H).
7. Secure supplies to effect cleanup.
8. During cleanup, wear appropriate apparel (see Sections I.F.1-3).
9. Notify the safety coordinator if a regulated substance is involved.

Handling of Spilled Liquids

1. Confine or contain the spill to a small area. Do not let it spread.
2. For small quantities of inorganic acids or bases, use a neutralizing agent or an absorbent mixture (e.g., soda ash or diatomaceous earth). For small quantities of other materials, absorb the spill with a nonreactive material (such as vermiculite, dry sand, or towels).
3. For larger amounts of inorganic acids and bases, flush with large amounts of water (provided that the water will not cause additional damage). Flooding is not recommended in storerooms where violent spattering may cause additional hazards or in areas where water-reactive chemicals may be present.
4. Mop up the spill, wringing out the mop in a sink or a pail equipped with rollers.
5. Carefully pick up and clean any cartons or bottles that have been splashed or immersed.
6. Vacuum the area with a vacuum cleaner approved for the material involved, remembering that the exhaust of a vacuum cleaner can create aerosols and, thus, should be vented to a hood or through a filter.
7. If the spilled material is extremely volatile, let it evaporate and be exhausted by the mechanical ventilation system (provided that the hood and associated mechanical system is spark-proof).
8. Dispose of residues according to safe disposal procedures (see Chapter II.G).

Handling of Spilled Solids

Generally, sweep spilled solids of low toxicity into a dust pan and place them in a solid-waste container for disposal. Additional precautions such as the use of a vacuum cleaner equipped with a HEPA filter may be necessary when cleaning up spills of more highly toxic solids.

HANDLING OF LEAKING COMPRESSED GAS CYLINDERS

Occasionally, a cylinder or one of its component parts develops a leak. Most such leaks occur at the top of the cylinder in areas such as the valve threads, safety device, valve stem, and valve outlet.

If a leak is suspected, do not use a flame for detection; rather, a flammable-gas leak detector or soapy water or other suitable solution should be used. If the leak cannot be remedied by tightening a valve gland or a packing nut, emergency action procedures should be effected and the supplier should be notified. Laboratory workers should never attempt to repair a leak at the valve threads or safety device; rather, they should consult with the supplier for instructions.

The following general procedures can be used for leaks of minimum size where the indicated action can be taken without serious exposure of personnel.

If it is necessary to move a leaking cylinder through populated portions of the building, place a plastic bag, rubber shroud, or similar device over the top and tape it (duct tape preferred) to the cylinder to confine the leaking gas.

1. Flammable, inert, or oxidizing gases—Move the cylinder to an isolated area (away from combustible material if the gas is flammable or an oxidizing agent) and post signs that describe the hazards and state warnings.
2. Corrosive gases may increase the size of the leak as they are released and some corrosives are also oxidants or flammable—Move the cylinder to an isolated, well-ventilated area and use suitable means to direct the gas into an appropriate chemical neutralizer. Post signs that describe the hazards and state warnings.
3. Toxic gases—Follow the same procedure as for corrosive gases. Move the cylinder to an isolated, well-ventilated area and use suitable means to direct the gas into an appropriate chemical neutralizer. Post signs that describe the hazards and state the warnings.

When the nature of the leaking gas or the size of the leak constitutes a more serious hazard, self-contained breathing apparatus (see Section I.F.4) or protective apparel (see Sections I.F.1-3), or both, may be required. Basic action for large or uncontrolled leaks may include any of the following steps:

1. evacuation of personnel,

2. rescue of injured personnel by crews equipped with adequate personal protective apparel and breathing apparatus,
3. fire fighting action,
4. emergency repair, and
5. decontamination.

II.F

Procedures for Special Disposal Problems that may Arise in Life Science Laboratories

The most appropriate method of disposing of animal tissues and excretion products that are contaminated with toxic substances and of the unused diet from studies in which the toxic substance was incorporated into the diet is by incineration (see also Section I.G.1).

Before incineration, however, it is important to consider the chemical properties and thermal stability, as well as the quantity, of the substance present in the items to be disposed of. For example, if the toxic substance in question is thermally stable, incineration in a conventional facility may lead to the contamination of the immediate environment of the incinerator. It is also important to know whether any product is formed on combustion of the animal tissues containing toxic substances that is thermally stable and likely to have significant toxicity. In rare cases, the combustion of animal tissues containing toxic substances may result in the formation of products that are more toxic than the material being disposed of. For example, recent work in Sweden has indicated that the incineration of polychlorinated biphenyls at 500-600°C may lead to the formation of highly toxic chlorinated benzofurans as a side product.

Thus, alternative methods of disposal should be used for the disposal of animal remains and products containing a thermally stable toxin or if it is anticipated that the combustion process will result in the formation of thermally stable toxins. One alternative that is available is combustion in a special incinerator capable of very high temperatures (1000°C) and having sufficient dwell time for the vaporized material to ensure its degradation.

Another, but less acceptable, alternative is burial of the animal remains and products in an EPA-approved burial site.

During the disposal process, care must be taken to package the animals and animal products in a way that minimizes potential exposure of all personnel to any toxic substances. Adequate protective apparel (see Sections I.F.1-3) and respirators (see Section I.F.4) (if needed) should be available.

II.G

Procedures for Disposal of Waste Materials from the Institution

Every organization should have a system for the disposal of chemical wastes that is safe and environmentally acceptable. This system should be under the direction of a safety coordinator or department specifically charged with that responsibility. Written plans for handling emergencies such as spills (see Section II.E.6) should be readily available. To serve the range of laboratory requirements (solid versus liquid, toxic versus nontoxic wastes), the disposal system should be complete and permit no loopholes. Storage of wastes at the institution should be minimal.

The Resources Conservation and Recovery Act (RCRA) of 1976, which is intended to provide cradle-to-grave control of hazardous waste, establishes a national hazardous-waste management program. The EPA is authorized to establish a system that will track hazardous chemical wastes from the time they are generated through their storage, transportation, and treatment to their ultimate disposal. The act requires that each step be documented, and regulations have been promulgated [45 Fed. Reg. 12,272-12,746 (Feb. 26, 1980) and 33,066-33,588 (May 19, 1980)] aimed at implementing this law.

It is now necessary to address the fundamental question of what constitutes a hazardous waste. The EPA proposal defines waste as hazardous if it meets criteria of toxicity, ignitability, corrosiveness, or reactivity. For example, ignitible waste is defined as (a) any liquid that has a flash point of less than 140°F, (b) any substance that can cause a fire by reaction or self-ignition, (c) any ignitable compressed gas, or (d) any oxidizer. [Lists of waste chemicals designated as hazardous are given elsewhere (40 CFR Part 250).]

Other regulations have been proposed concerning the generation and transportation of chemical wastes, waste-facility standards, permits, state programs, and inspection and enforcement. Implementation, however, will occur one step at a time; thus, disposal procedures should be confirmed with a knowledgeable safety coordinator at least annually. The RCRA will have a major impact on all aspects of chemical waste disposal.

II.G.1 INCINERATION AND SOLVENT BURNERS

Incineration is the most environmentally acceptable method of chemical waste disposal. Combustion of organic materials with excess oxygen at high temperatures for sufficient time results in degradation to elemental constituents or by-products that are easier to handle in an environmentally acceptable manner. In addition to heat, the principal products of incineration are carbon dioxide, water, and oxides of sulfur and nitrogen; depending on what is burned, other volatile materials may also be formed. Nonvolatile products include fly ash and solid residues.

Incinerator technology is highly developed; a wide variety of sizes and types of equipment that can be used to handle solids, liquids, and gases is available. Current regulations mandate that such units be equipped with secondary-treatment devices such as afterburners, scrubbers, electrostatic precipitators, and filters. Incinerators are complex pieces of equipment that require competent operating personnel as well as operating permits. Some incinerators can be equipped to burn discarded solvents for their heat value. (See also *Cleaning Our Environment: A Chemical Perspective*, 2nd ed.; American Chemical Society: Washington, D.C., 1978, and Powers, W. *How to Dispose of Toxic Substances and Industrial Wastes*, Noyes Data Corp.: Park Ridge, N.J., 1976.)

The EPA, under the RCRA, has promulgated regulations setting performance standards for incinerators. If not modified before becoming final, these regulations will require destruction combustion efficiencies of 99.9% and set particulate limits and retention times at specified temperatures.

Organizations that do not have access to an incinerator can arrange for the collection and incineration of chemical wastes by contract with a supplier of such services. However, the transportation of hazardous chemicals must be in accordance with the DOT regulations (49 CFR Parts 100-199).

II.G.2 SEWERS

Many water-soluble chemicals can safely be flushed down the drain into sewers that go to treatment plants. However, care must be taken to avoid using this disposal system for materials that can create problems in the sewer system or constitute a violation of regulations. In the United States, local municipal codes and regulations govern discharge into public sanitary and storm drain systems and usually have limitations regarding the introduction of chemical wastes into these systems. It is essential, therefore, to be aware of these limitations and to obtain any necessary permits.

There are also EPA-promulgated regulations on prohibited discharges in the general pretreatment regulations for existing and new sources of pollution as outlined below. States and municipalities will extend the regulation of sewer discharges as these standards are implemented.

1. Pollutants that inhibit or interfere with the operation or performance of the publicly owned treatment works (POTW) shall not be introduced.
2. The following pollutants may not be introduced: (a) those that can create a fire or explosion hazard, (b) those that can cause structural damage (no discharge at a pH lower than 3 unless the POTW is designed to handle it, (c) solid or viscous materials in amounts that could obstruct the flow or interfere with the operation of the POTW, (d) those released in such volume or strength as to cause interference with operation of the POTW, and (e) heat in amounts that could inhibit biological activity in the POTW.
3. Those POTW developing pretreatment programs must develop and enforce specific limits for the above pollutants. In addition, programs by a POTW pursuant to obtaining a permit under the National Pollutant Discharge Elimination Program may require the development and setting of specific limits.

It should be noted that these limitations do not exempt material that might be discharged because of an accidental spill. Therefore, it is necessary to provide for control of discharges of spilled substances by, for example, special trapped floor drains that discharge to a holding tank or by an environmentally safe drainage and treatment facility.

II.G.3 LANDFILLS

Landfill or burial is a common method for disposal of chemical wastes; both public and private landfills are important outlets for such wastes.

This disposal procedure, however, has led to a dispersal of wastes in the environment and is often only a postponement of the ultimate problem. Local ordinances should be checked, and all disposal practices should be in compliance with federal regulations; great care should be taken to ensure that potentially hazardous chemicals are treated or disposed of in a manner that avoids personal or environmental risks either currently or in the future.

Many toxic chemicals can be rendered innocuous for landfill disposal by oxidation, reduction, or complexation. Special chemical or secure landfills are also available for disposing of potentially injurious chemicals.

The EPA has proposed, under RCRA, regulations for chemical waste disposal via landfill [45 Fed. Reg. 33,209-33,215 (May 19, 1980)]. The date of implementation will be variable depending on the degree of hazard, but the regulations are comprehensive, covering the construction, operation, and monitoring of a landfill. Violators will be subject to severe fines.

In addition, many states have adopted their own set of disposal regulations and local administrators should be consulted for guidance and assistance.

Organizations that contract for waste disposal should ensure that the contractor is performing in a safe, legal, and responsible manner.

II.G.4 RECYCLING

Because of the constraints being placed on chemical waste disposal, the cost of acceptable disposal methods is increasing dramatically. This process will result in increased emphasis on recovery and recycling of chemicals that were formerly discarded. Organizations should consider establishing procedures for this purpose depending on the types of materials used and their suitability for recovery operations. For example, many laboratories have a system for the recovery and reuse of mercury.

II.G.5 COMPRESSED GAS CYLINDER VENTING

Occasionally, it is necessary to vent leaking or damaged cylinders of compressed gases (see Section II.E.6). Such gases may be toxic, flammable, or corrosive. This procedure should be carried out by personnel trained in the safe handling of compressed gas cylinders. Safety equipment such as respirators (see Section I.F.4) should be available, and adequate ventilation to avoid personnel exposure or explosion should be provided. Disposal facilities equipped with scrubbing or incinerating devices may be required depending on such factors as the volume and nature of the material, the physical location, and local regulations. Venting should be done at a rate that will avoid environmental and safety problems and in compliance with local emission regulations.

II.G.6 TRANSPORT OF CHEMICAL WASTES FROM THE INSTITUTION

Adherence to the DOT regulations regarding the transportation of hazardous substances must be ensured. Laboratory workers should know that no quantity of chemically hazardous material may be transported on any public conveyance, by the U.S. Postal Service, or by a private conveyance over public highways without proper packaging classification, labeling, and documentation.

APPENDIXES

Appendix A

Evaluation of Published Epidemiological Studies of Chemists and Recommendations Concerning Future Studies

Several epidemiological studies of relative mortality experience among chemists have been reported (*1-6*). Until recently, study results (*1-5*) consistently showed a higher than expected risk of death from cancer. This evidence provided some concern over long-term effects imputed to the environment to which this particular occupational group is exposed.

The most recent evidence (*6*) is at variance with the previously reported findings. However, inconsistency in the results of the various epidemiological studies is not unexpected due to differences in design.

In evaluating the evidence and conclusions of these epidemiological studies, the following characteristics are reviewed:

1. the study methodology used;
2. the specificity of the exposures, i.e., evidence on the work environments of those in the various study groups;
3. the composition of the study populations; and
4. the consistency of the results when the several studies, and other evidence, are examined.

This review attempts a critical appraisal of the published evidence. The appraisal, in turn, suggests selecting one among several possible conclusions: Is evidence present or not present and, if present, how strong is the evidence suggesting increased risk? Finally, brief consideration is given to alternative study designs that might elucidate further any increased risk of cancer for those in the work environment of the laboratory. The latter are proposed, in part, based on this appraisal of the published studies.

1 CRITIQUE

METHODOLOGY

A study by Li *et al.* (*1*) has been cited frequently as suggesting an increased risk of death from cancer among chemists. The data base for this study was derived from deaths among American Chemical Society (ACS) members reported in *Chemical and Engineering News* between April 1948 and July 1967. Relative frequencies of cause-specific mortality for those aged 20-64 years (divided into four age subgroupings) were compared with those of a sample of professional men in the same age groups whose deaths occurred in 1950. For those deaths to males at ages 65 and over, the comparison group of U.S. white male mortality was used. Comparison data for females were derived from published mortality information on the U.S. white female population in 1959. Cause of death and other vital data about the ACS group were obtained from death certificates. For 4611 death notifications, 3637 certificates were obtained, a 78% completion proportion.

Li *et al.*, themselves, while concluding that "the study group had a significantly increased proportion of deaths attributed to cancer, particularly of the pancreas and lymphoid tissue, a finding which suggests the influence of carcinogens encountered by chemists in their work," also pointed out that there are methodological problems in their study.

Thus, cautions in interpreting the Li *et al.* study include the following:

1. A higher proportion of cancer deaths among ACS members was associated with a finding of lower relative frequency (expected compared with observed) of deaths from other causes. This observation reinforces concern about interpreting findings derived by the relative-frequency method. For example, should ACS members have more favorable longevity in contrast to a comparison group, an excess of cause-specific deaths could be shown, but might be an artifact of the method used. In fact, both study and comparison groups could have the same true frequency of cause-specific mortality. Other methods of study would be needed to confirm or refute such an hypothesis.

2. The comparison data varied in two respects from the study data. First, ACS member deaths were accumulated over an approximately 20-year period, whereas the comparison data were based on a sample of deaths among U.S. professional men dying in 1950, both groups aged 20-

64 years. Changes in the incidence of cancer over time, total as well as site- and age-specific, could bias the findings as different time periods were compared. Second, all deaths among white U.S. males aged 65 and over were compared with deaths among ACS members at the same ages. Mortality in this general U.S. population may well be different from that observed in a group of members of a professional society.

3. Cautious interpretation of the study findings is required because of

a. incompleteness in tracing all deaths notified,

b. possible preferential reporting of deaths due to cancer as compared with deaths from other causes, and

c. chance or other selection biases that may influence ACS membership and mortality.

Olin (*2-3*) studied deaths among graduates of the years 1930-1950 from the two major chemical engineering training institutions in Sweden. The first report (*2*) reviewed the mortality experience of 530 males who graduated from the School of Chemical Engineering at the Royal Institute of Technology, Stockholm. Death certificates were used as the source document for the 58 who were deceased among the 517 graduates who could be traced. In a more recent article (*3*), 335 male graduates of the Chalmers Institute of Technology, Gothenburg, were added to the original study population. The combined group of 857 were traced, with an approximately 99% completion proportion. A total of 93 deaths were discovered. Expected numbers of cause-specific deaths were estimated in 5-year age ranges, based on Swedish mortality data from 1973, and compared with the observed deaths (similar age groupings). An excess of malignant tumors (34) was found in comparison with the expected number (20.8). The increase was due in part to an excess of leukemia and malignant lymphoma deaths, three of which were classified as Hodgkins disease. An excess of prostatic cancer was also found.

A faculty member at the Royal Institute of Technology was reported to have distinguished between graduates who had worked in the laboratory ("chemists") and those who had not ("nonchemists"). Olin observed, "all but one of the 22 cancer deaths occurred in the chemists group . . . this result strongly suggests that the difference in the neoplasm death rates of the two groups is at least partly attributable to work in chemical laboratories" (*2*). In addition, Olin noted that eight of nine chemists who died of leukemia/lymphoma or neoplasms of the urinary organs were classified as organic chemists. No classification of type of work, nor the mortality experience, was reported for the graduates of the Chalmers Institute of Technology.

The study groups reported in the Olin papers, and the methodology used in these studies, present interpretative difficulties similar to those encountered in the Li *et al.* study. Cautions in interpreting Olin's results include the following:

1. The death certificate served as the basis for cause-specific mortality comparisons. It is not clear from the papers how coding of cause of death was performed, i.e., standardized according to internationally accepted coding criteria.
2. Expected cause-specific deaths were generated from general mortality statistics for 1973. Observed deaths were accumulated over a follow-up period extending from a date of graduation falling between the years 1930 and 1950 and the completion of the follow-up phase of the study, December 31, 1974.
3. Comparisons based on general population mortality must be interpreted with caution, particularly when the group of deaths observed in the study population derives from an employed group, probably predominantly residing in urban areas and having income and education levels suggestive of higher average socioeconomic status than the general population.
4. Exposure information may have been collected in a somewhat nonstandardized form. Classifications of employment described, chemist and nonchemist, indicate only general work circumstances. Reported findings of excess leukemia/lymphoma and urinary tract cancer deaths among those classified as organic chemists must be interpreted cautiously. No information is given on the division of chemists among the branches of that particular science.
5. The numbers of cases, both observed and expected, were small. The higher risk of death from cancer reported was based on proportionate mortality analysis. Standardized mortality for the study population, when compared with the general population, showed a slightly lower risk of cancer deaths for the chemists.

In an abstract, Searle *et al.* (*4*) reported an analysis of deaths among 1332 members of the Royal Institute of Chemistry who died in the years 1965-1975. An excess of deaths from cancer is shown. These authors conclude that "though analysis of deaths is in an early stage, it appears that the incidence of lymphomas may be elevated in the RIC members also."

McGinty (*7*) has also reported on the study population described in the Searle *et al.* abstact. McGinty concludes "For example, the death rate from cancer among chemists is, it appears, distinctly lower than would be expected." He went on to report a slight excess of lymphomas and chronic lymphatic leukemia deaths and a two to four times excess mortality above expected for deaths from cancer of the small and large intestine. The McGinty paper, however, presented no data.

Milham (*5*) has summarized occupational mortality as reported on death certificates for Caucasian males 20 years of age and older in the state of Washington, during 1950-1971. Proportionate mortality ratios (PMRs) were used, and the data were programmed such that, for each occupational category examined, deaths observed in excess of those expected were noted. For the occupational category of "chemists," 172 deaths were recorded. PMRs in excess of expected were observed for a number of specific causes of death, including cancer (pancreas and lymphatic and hematopoeitic tissues as well as leukemia and aleukemia). By statistical test, only suicide and cerebral embolism and thrombosis were found significantly in excess.

Hoar (*6*) studied cancer incidence and absolute mortality among chemists employed by a single large company (E.I. du Pont de Nemours & Co.). A cohort of 3686 male and 75 white female chemists and persons in related occupational groups (e.g., chemical engineers) was identified from a file of all salaried personnel actively employed on or after January 1, 1964, and for whom work history and other information was available to 1959. A comparison group of 19,262 male and 673 female white "nonchemists" was identified from the same file.

Cancer cases were identified from a company-maintained registry that included notifications of causes of disability and death from accident and health insurance, as well as life insurance claims. Cancer incidence information was available only for those actively employed over the study interval, 1964-1977.

Cancer incidence among the group of employees characterized as "chemists," among the nonchemists comparison group, and derived from the Third National Cancer Survey were calculated as standardized incidence ratios (SIRs).

Cancer mortality experience for both the chemist and the nonchemist groups was derived from company records and the Social Security Administration for all in the cohort (regardless of whether or not an employee at the close of the study period, December 31, 1977). Death certificates were obtained in high proportion for those known to have died in the United States. Standardized mortality ratios (SMRs) were calculated from the data derived on the two employee groups and for the U.S. general population.

Among the 3686 white male chemists observed, 61 cases of cancer were diagnosed between 1964 and 1977. An expected 86.5 cases, based on the nonchemist experience, gave a SIR of 71 for the chemists. Comparative cancer incidence data from the Third National Cancer Study yielded a SIR of 54 (112.0 cases expected). The lower cancer incidence among the chemists was due in part to a deficit in lung cancer cases (8 observed, 20 expected). Although numbers were small, cases of pancreatic, prostatic, urinary bladder, and lymphatic and hematopoeitic system neoplasms were observed. Comparisons of this cancer incidence with both nonchemist and Third National Cancer Survey data were consistent, except for a higher SIR of melanoma in both employee groups in contrast to the general population incidence data.

Cancer mortality experience among the chemists was lower than that observed for the nonchemists (SMR 69). As was noted in the comparisons of incidence, a lower mortality from lung cancer was observed among the chemists, as well as larger than expected numbers of several site-specific neoplasms (Hodgkins and all lymphatic and hematopoeitic). A significant excess of cancer of the large intestine was found.

The results of this study are at variance with other reports, a point noted above. These results were derived by methods generally not used in the other investigations. The methodological differeces include the following:

1. Completed or absolute cancer mortality (SMR) versus incidence (SIR) measures—only the Olin (*2, 3*) study used absolute-mortality methodology, but the resulting numbers of cases were generally too small for complete analysis.
2. Emphasis on comparing mortality and incidence between a population of chemists and a parallel group drawn from a similar work-and-living environment, the differentiating variable presumably being the specific work setting (microenvironment) of the chemist population.

The importance of interpreting the findings reported by Hoar (*6*) in light of the methodology used is emphasized by comparing the PMRs with the SMRs derived from the data set. These are (PMR versus SMR), for all cancers, 116 to 50; for cancer of the large intestine, 426 to 174; and for leukemia, 230 to 119.

The results of the Hoar study, at variance with findings in several of the other studies, should be interpreted with care because of the following possible confounding factors:

1. Although in this investigation chemist and nonchemist employees of the same company were identified, chemical exposure histories were not available. For instance, a subset of the chemists may have had exposures that resulted in developing a cancer, but because of the small numbers, this risk is obscured.

2. Both employee groups, chemist and nonchemist, include individuals who have had job mobility within the single company. Some potential for misclassification must be accepted.

3. At the close of the period of observation (December 31, 1977), approximately three-quarters of the chemists were still actively employed. Cancer deaths occurring later in life would be detected only by a longer follow-up period. An excess of cancer deaths might then be observed among the chemists. Comparative differences in mortality ratios would thus be reduced.

4. The two employee groups, chemists and nonchemists, may differ in ways that are associated with differential risk of developing cancer. For example, indirect evidence suggests that cigarette smoking habits vary between the two groups.

5. The results of this study derive from observing employee groups of a single company. Generalizing the findings to all chemists requires strong assumptions about similarities of circumstances in the workplace of chemists.

In summary, the results of the Hoar study provide a perspective from which to evaluate the findings of the other studies; the earlier and consistent findings of the higher risk of cancer among those identified variously as chemists is not confirmed. In large measure, the difference lies in the methodology used. Confirmation of these findings by other studies using similar methods, as well as longer-term follow-up of cohorts now identified and having sufficient numbers for analysis, could provide further evidence on which to base decisions about cancer risk to chemists.

Exposure Data

Several of those whose work has been reviewed have pointed out the absence of exposure data among the study group, as well as the inherent difficulties in obtaining such data. Unless, fortuitously, very good exposure information had been accumulated over a sufficient length of time and in a consistent fashion, it would be extremely difficult (if not impossible) to reconstruct exposure histories for those who were deceased. Methods of classifying exposure as reported by Olin have been commented on above. Both McGinty and Searle *et al.* indicate that some exposure information for members of the Royal Institute of Chemistry may provide the basis for reevaluation of retrospective data and the development of prospective studies. It is not clear from either paper whether exposure data will be sufficiently refined to indicate "microenvironments" in which individuals have worked, including type and amount as well as duration of exposures.

Study Populations

The definition of who has been included, and who has not been included, in the occupational category of chemist is a problem to some degree in all of the studies examined. Li *et al.* based their observations on deaths among members of the ACS, those holding membership during an approximately 20-year time span and whose deaths were notified to the Society and published in *Chemical and Engineering News.* Olin used a relatively complete follow-up of persons who had graduated from the two principal institutions for the training of chemists in Sweden. The data base for the Searle *et al.* study consisted of members of the Royal Institute of Chemistry of Great Britain. McGinty, in reporting on mortality ratios among chemists in Britain, pointed out that the occupational classifications used excluded a number of persons involved in chemistry-type activities, e.g., teachers, chemical engineers, and technicians. In addition, the occupational category used was the most recent occupation. The Milham study used death certificates; PMRs were calculated for occupational groupings according to the reported "usual occupation during most of lifetime" as shown there.

Membership in a professional society is, in large measure, a choice made by an individual who is eligible to join. Society membership may be assumed to be inclusive of the entire occupational or professional group eligible and at risk. A decision on whether or not to take membership may be biased in a number of directions, many of those directions discoverable, if at all, only by extensive sampling, inquiry, and further analysis.

Consistency of Results

The Li *et al.*, Olin, Searle *et al.*, and Milham reports are consistent, given methodological limitations, in the finding of excess cancer mortality among persons classified as chemists. All three of these studies used PMR analysis in reaching their conclusions. The results reported by Olin showed excess cancers in chemists by PMR but no excess when absolute mortality was analyzed. Similarly, McGinty, using British national mortality data for 1970-1972, found an SMR for chemists of 89. The study by Hoar likewise showed a lower than expected absolute mortality (SMR) and incidence (SIR) of chemists when contrasted with the experience of several comparison populations. Evidence is, thus, inconsistent on whether or not chemists have a higher risk of cancer. The inconsistency results from use of different methodologies in the several studies.

Consistency of findings among the studies of excess risk for type-specific cancers is variable. A large number of specific cancer types are reported as in excess of expected. However the small numbers reported confound the analysis, and the bias introduced by the use of proportionate mortality analysis further confounds any interpretation of these findings.

2 CONCLUSIONS

As discussed above, whether any increased risk of death from cancer exists for those variously classified in the occupational category of chemist has not been consistently shown in studies reported to date. Although some investigations, particularly those using proportionate mortality analysis, find evidence of higher risk, other studies, those using standardized mortality approaches, do not confirm these findings. Similarly, interpretation of data on type-specific cancer excess is complicated. Findings of observed cases higher than expected in a particular study are generally characterized by small numbers. Other methodological problems confounding the interpretation of these findings have been discussed above.

3 FUTURE STUDIES

A critical review of the published studies of cancer mortality among chemists indicates that future studies should include the following design characteristics:

1. Identification of a group of chemists should be inclusive and representative of the universe of chemists. Society membership, for example, is inadequate for defining a cohort for study due to the unknown variables governing the decision of whether or not an individual joins a professional association.
2. A history of exposure is needed for each member of the group under study. It seems unlikely that proxies, such as the branch of chemistry in which a particular person is engaged, will provide other than a gross approximation to the extent and nature of chemical exposures. Whether a valid instrument or methodology for accumulating an exposure history can be devised remains a difficult, but fundamental, further problem for study design.

The various surveillance, epidemiology, and end results programs, particularly those registries of all cancer cases occurring within large, defined geographical areas, provide a potential data base for future cohort studies. For example, should a cohort of chemists be identified within an area served by such a registry, cancer incidence could be ascertained during the follow-up period by using, in part, case data within the registry.

Long-term study of chemists and a variety of comparison populations could be developed by using various industrial and other occupational populations. Possibilities of defining and recording work-related exposure histories would appear to be enhanced within the current concerns over workplace safety. The importance, in such studies, of careful definition of the occupational groups studied and of maintaining accurate exposure histories has been discussed above. The need for accumulating sufficiently large numbers of observations over long time periods and for the use of the most appropriate analytical methods has been stressed. If sufficiently large numbers of cancer cases among cohorts of chemists accumulate, case control studies might be used. This analytical approach could provide information on the association between occupational exposures and the presence of type-specific neoplasms.

4 REFERENCES

1. Li, I.; Fraumeni, J.; Mantel, N.; Miller, R. *Cancer Mortality among Chemists*; J. Natl. Cancer Inst., 1969, Vol. 43, pp. 1159-64.

2. Olin, R. *Leukemia and Hodgkin's Disease among Swedish Chemistry Graduates*; Lancet (ii), 1976, p. 916.

3. Olin, R. *The Hazards of a Chemical Laboratory Environment: A Study of the Mortality in two Cohorts of Swedish Chemists*; Amer. Ind. Hygiene Assn. J., 1978, Vol. 39, pp. 557-62.

4. Searle, C.; Waterhouse, J.; Henman, B.; Bartlett, D.; McCombie, S. *Epidemiological Study of the Mortality of British Chemists*. Brit. J. Cancer, 1978, Vol. 38, pp. 192-93.

5. Milham, S. Occupational Mortality in Washington State, 1950-1971. Department of Health, Education and Welfare, Rept. NIOSH 76-175-B. U.S. Government Printing Office, Washington, D.C., 1976.

6. Hoar, S. *A Retrospective Cohort Study of Chemists Employed by E.I. du Pont de Nemours and Company, Inc.*; Department of Epidemiology, Harvard University School of Public Health, Boston, Mass.; D. Sci. thesis, 1980.

7. McGinty, L. Not So Grim Reaper. Chem. in Brit., 1978, Vol. 14, pp. 508, 511-12, 514.

Appendix B

TLVs®

Threshold Limit Values for Chemical Substances and Physical Agents in Workroom Air Adopted by ACGIH for 1980*

*Reproduced with permission from a booklet published by the American Conference of Governmental Hygienists, Inc., Cincinnati, Ohio.

PREFACE
CHEMICAL CONTAMINANTS

Threshold limit values refer to airborne concentrations of substances and represent conditions under which it is believed that nearly all workers may be repeatedly exposed day after day without adverse effect. Because of wide variation in individual susceptibility, however, a small percentage of workers may experience discomfort from some substances at concentrations at or below the threshold limit; a smaller percentage may be affected more seriously by aggravation of a pre-existing condition or by development of an occupational illness.

Tests are available (J. Occup. Med. 15: 564, 1973; Ann. N.Y. Acad. Sci., *151, Art. 2:* 968, 1968) that may be used to detect those individuals hypersusceptible to a variety of industrial chemicals (respiratory irritants, hemolytic chemicals, organic isocyanates, carbon disulfide).

Three categories of Threshold Limit Values (TLVs) are specified herein, as follows:

a) Threshold Limit Value-Time Weighted Average (TLV-TWA) — the time-weighted average concentration for a normal 8-hour workday or 40-hour workweek, to which nearly all workers may be repeatedly exposed, day after day, without adverse effect.

b) Threshold Limit Value-Short Term Exposure Limit (TLV-STEL) — the maximal concentration to which workers can be exposed for a period up to 15 minutes continuously without suffering from 1) irritation, 2) chronic or irreversible tissue change, or 3) narcosis of sufficient degree to increase accident proneness, impair self-rescue, or materially reduce work efficiency, provided that no more than four excursions per day are permitted, with at least 60 minutes between exposure periods, and provided that the daily TLV-TWA also is not exceeded. The STEL should be considered a maximal allowable concentration, or ceiling, not to be exceeded at any time during the 15-minute excursion period. STELs are based on one or more of the following criteria: (1) Adopted TLVs including those with a "C" or "ceiling" limit. (2) Pennsylvania Short-Term Limits for Exposure to Airborne Contaminants (Penna. Dept. of Hlth., Chapter 4, Art. 432, Rev. Jan. 25, 1968). (3) OSHA Occupational Safety and Health Standards, 40 FR 23073, May 28,

1975. (4) NIOSH criteria for recommended standards for occupational exposure to specific substances. The TWA-STEL should not be used as engineering design criterion or considered as an emergency exposure level (EEL).

c) Threshold Limit Value-Ceiling (TLV-C) — the concentration that should not be exceeded even instantaneously.

For some substances, e.g., irritant gases, only one category, the TLV-Ceiling, may be relevant. For other substances, either two or three categories may be relevant, depending upon their physiologic action. It is important to observe that if any one of these three TLVs is exceeded, a potential hazard from that substance is presumed to exist.

The TLV-TWA should be used as guides in the control of health hazards and should not be used as fine lines between safe and dangerous concentrations.

Time-weighted averages permit excursions above the limit provided they are compensated by equivalent excursions below the limit during the workday. In some instances it may be permissible to calculate the average concentration for a workweek rather than for a workday. The relationship between threshold limit and permissible excursion is a rule of thumb and in certain cases may not apply. The amount by which threshold limits may be exceeded. for short periods without injury to health depends upon a number of factors such as the nature of the contaminant, whether very high concentrations — even for short periods — produce acute poisoning, whether the effects are cumulative, the frequency with which high concentrations occur, and the duration of such periods. All-factors must be taken into consideration in arriving at a decision as to whether a hazardous condition exists.

Threshold limits are based on the best available information from industrial experience, from experimental human and animal studies, and, when possible, from a combination of the three. The basis on which the values are established may differ from substance to substance; protection against impairment of health may be a guiding factor for some, whereas reasonable freedom from irritation, narcosis, nuisance or other forms of stress may form the basis for others.

The amount and nature of the information available for establishing a TLV varies from substance to substance; consequently, the precision of the estimated TLV is also subject to variation and the latest

Documentation should be consulted in order to assess the extent of the data available for a given substance.

The committee holds to the opinion that limits based on physical irritation should be considered no less binding than those based on physical impairment. There is increasing evidence that physical irritation may initiate, promote or accelerate physical impairment through interaction with other chemical or biologic agents.

In spite of the fact that serious injury is not believed likely as a result of exposure to the threshold limit concentrations, the best practice is to maintain concentrations of all atmospheric contaminants as low as is practical.

These limits are intended for use in the practice of industrial hygiene and should be interpreted and applied only by a person trained in this discipline. They are not intended for use, or for modification for use, (1) as a relative index of hazard or toxicity, (2) in the evaluation or control of community air pollution nuisances, (3) in estimating the toxic potential of continuous, uninterrupted exposures or other extended work periods, (4) as proof or disproof of an existing disease or physical condition, or (5) for adoption by countries whose working conditions differ from those in the United States of America and where substances and processes differ.

Ceiling vs Time-Weighted Average Limits. Although the time-weighted average concentration provides the most satisfactory, practical way of monitoring airborne agents for compliance with the limits, there are certain substances for which it is inappropriate. In the latter group are substances which are predominantly fast acting and whose threshold limit is more appropriately based on this particular response. Substances with this type of response are best controlled by a ceiling "C" limit that should not be exceeded. It is implicit in these definitions that the manner of sampling to determine noncompliance with the limits for each group must differ; a single brief sample, that is applicable to a "C" limit, is not appropriate to the time-weighted limit; here, a sufficient number of samples are needed to permit a time-weighted average concentration throughout a complete cycle of operations or throughout the work shift.

Whereas the ceiling limit places a definite boundary which concentrations should not be permitted to exceed, the time-weighted average limit requires an explicit limit to the excursions that are permissible above the listed values. It should be noted that the same factors are used by the Committee in determining the magnitude of the value of the STELs, or whether to include or exclude a substance for a "C" listing.

"Skin" Notation. Listed substances followed by the designation "Skin" refer to the potential contribution to the overall exposure by the cutaneous route including mucous membranes and eye, either by airborne, or more particularly, by direct contact with the substance. Vehicles can alter skin absorption. This attention-calling designation is intended to suggest appropriate measures for the prevention of cutaneous absorption so that the threshold limit is not invalidated.

Mixtures. Special consideration should be given also to the application of the TLVs in assessing the health hazards which may be associated with exposure to mixtures of two or more substances. A brief discussion of basic considerations involved in developing threshold limit values for mixtures, and methods for their development, amplified by specific examples are given in Appendix C.

Nuisance Particulates. In contrast to fibrogenic dusts which cause scar tissue to be formed in lungs when inhaled in excessive amounts, so-called "nuisance" dusts have a long history of little adverse effect on lungs and do not produce significant organic disease or toxic effect when exposures are kept under reasonable control. The nuisance dusts have also been called (biologically) "inert" dusts, but the latter term is inappropriate to the extent that there is no dust which does not evoke some cellular response in the lung when inhaled in sufficient amount. However, the lung-tissue reaction caused by inhalation of nuisance dusts has the following characteristics: (1) The architecture of the air spaces remains intact. (2) Collagen (scar tissue) is not formed to a significant extent. (3) The tissue reaction is potentially reversible.

Excessive concentrations of nuisance dusts in the workroom air may seriously reduce visibility, may cause unpleasant deposits in the eyes, ears and nasal passages (Portland Cement dust), or cause injury to the skin or mucous membranes by chemical or mechanical action per se or by the rigorous skin cleansing procedures necessary for their removal.

A threshold limit of 10 mg/m^3, or 30 mppcf, of total dust < 1% quartz, or, 5 mg/m^3 respirable dust is recommended for substances in these categories and for which no specific threshold limits have been assigned. This limit, for a normal workday, does not apply to brief exposures at higher concentrations.

Neither does it apply to those substances which may cause physiologic impairment at lower concentrations but for which a threshold limit has not yet been adopted. Some nuisance particulates are given in Appendix E.

Simple Asphyxiants — "Inert" Gases or Vapors. A number of gases and vapors, when present in high concentrations in air, act primarily as simple asphyxiants without other significant physiologic effects. A TLV may not be recommended for each simple asphyxiant because the limiting factor is the available oxygen. The minimal oxygen content should be 18 percent by volume under normal atmospheric pressure (equivalent to a partial pressure, pO_2 of 135 mm Hg). Atmospheres deficient in O_2 do not provide adequate warning and most simple asphyxiants are odorless. Several simple asphyxiants present an explosion hazard. Account should be taken of this factor in limiting the concentration of the asphyxiant. Specific examples are listed in Appendix F.

Physical Factors. It is recognized that such physical factors as heat, ultraviolet and ionizing radiation, humidity, abnormal pressure (altitude) and the like may place added stress on the body so that the effects from exposure at a threshold limit may be altered. Most of these stresses act adversely to increase the toxic response of a substance. Although most threshold limits have built-in safety factors to guard against adverse effects to moderate deviations from normal environments, the safety factors of most substances are not of such a magnitude as to take care of gross deviations. For example, continuous work at temperatures above 90°F, or overtime extending the workweek more than 25%, might be considered gross deviations. In such instances judgment must be exercised in the proper adjustments of the Threshold Limit Values.

Biologic Limit Values (BLVs). Other means exist and may be necessary for monitoring worker exposure other than reliance on the Threshold Limit Values for industrial air, namely, the Biologic Limit Values. These values represent limiting amounts of substances (or their effects) to which the worker may be exposed without hazard to health or well-being as determined in his tissues and fluids or in his exhaled breath. The biologic measurements on which the BLVs are based can furnish two kinds of information useful in the control of worker exposure: (1) measure of the individual worker's over-all exposure; (2) measure of the worker's individual and characteristic response. Measurements of response furnish a superior

estimate of the physiologic status of the worker, and may be made of (a) changes in amount of some critical biochemical constituent, (b) changes in activity of a critical enzyme, (c) changes in some physiologic function. Measurement of exposure may be made by (1) determining in blood, urine, hair, nails, in body tissues and fluids, the amount of substance to which the worker was exposed; (2) determination of the amount of the metabolite(s) of the substance in tissues and fluids; (3) determination of the amount of the substance in the exhaled breath. The biologic limits may be used as an adjunct to the TLVs for air, or in place of them. The BLVs, and their associated procedures for determining compliance with them, should thus be regarded as an effective means of providing health surveillance of the worker.

Unlisted Substances. Many substances present or handled in industrial processes do not appear on the TLV list. In a number of instances the material is rarely present as a particulate, vapor or other airborne contaminant, and a TLV is not necessary. In other cases sufficient information to warrant development of a TLV, even on a tentative basis, is not available to the Committee. Other substances, of low toxicity, could be included in Appendix E pertaining to nuisance particulates. This list (as well as Appendix F) is not meant to be all inclusive; the substances serve only as examples.

In addition there are some substances of not inconsiderable toxicity, which have been omitted primarily because only a limited number of workers (e.g., employees of a single plant) are known to have potential exposure to possibly harmful concentrations.

"Notice of Intent." At the beginning of each year, proposed actions of the Committee for the forthcoming year are issued in the form of a "Notice of Intended Changes." This Notice provides not only an opportunity for comment, but solicits suggestions of substances to be added to the list. The suggestions should be accompanied by substantiating evidence. The list of Intended Changes follows the Adopted Values in the TLV booklet. Values listed in parenthesis in the "Adopted" list are to be used during the period in which a proposed change for that Value is listed in the Notice of Intended Changes.

Legal Status. The Threshold Limit Values, as issued by ACGIH, are recommendations and should be used as guidelines for good practices. Wherever these values (of whatever year) have been used or included by reference in Federal and/or State statutes and registers, the TLVs do have the force and effects of law.

Reprint Permission. This book is fully protected by copyright and no part of it may be reproduced in any form, by print, photoprint, microfilm or any other means without written permission from the American Conference of Governmental Industrial Hygienists, Inc., P.O. Box 1937, Cincinnati, OH 45201.

Substance	ADOPTED VALUES		TENTATIVE VALUES	
	TWA		STEL	
	ppm[a]	mg/m³[b]	ppm[a]	mg/m³[b]
Abate	—	10	—	20
Acetaldehyde	100	180	150	270
Acetic acid	10	25	15	37
C Acetic anhydride	5	20	—	—
** Acetone	(1,000)	(2,400)	(1,250)	(3,000)
Acetonitrile	40	70	60	105
Acetylene	F	—	—	—
Acetylene dichloride, see 1, 2-Dichloroethylene	200	790	250	1,000
Acetylene tetrabromide	1	15	1.5	20
* Acetylsalicylic acid (Asprin)	—	5	—	—
Acrolein	0.1	0.25	0.3	0.8
Acrylamide — Skin	—	0.3	—	0.6
** Acrylonitrile — Skin	(A1b)	(A1b)	—	—
Aldrin — Skin	—	0.25	—	0.75
Allyl alcohol — Skin	2	5	4	10
Allyl chloride	1	3	2	6
Allyl glycidyl ether (AGE) — Skin	5	22	10	44
Allyl propyl disulfide	2	12	3	18
Aluminum metal and oxide	—	10	—	20
Aluminum pyro powders	—	5	—	—
Aluminum welding fumes	—	5	—	—
Aluminum, soluble salts	—	2	—	—
Aluminum, alkyls (NOC)*	—	2	—	—
Aluminum oxide (Al_2O_3)	—	E	—	20
4-Aminodiphenyl — Skin	—	A1b	—	A1b
2-Aminoethanol, see Ethanolamine	3	8	6	15
2-Aminopyridine	0.5	2	2	4
3-Amino 1, 2, 4-triazole	A2	A2	—	—
Ammonia	25	18	35	27
Ammonium chloride-fume	—	10	—	20
Ammonium sulfamate (Ammate)	—	10	–	20
n-Amyl acetate	100	530	150	800
sec-Amyl acetate	125	670	150	800
* Aniline & homologues — Skin	2	10	5	20

Capital letters refer to Appendices.
Footnotes (a thru f) see Page 32.
*1980 Addition
**See Notices of Intended Changes

Substance	ADOPTED VALUES		TENTATIVE VALUES	
	TWA		STEL	
	ppm[a]	mg/m³[b]	ppm[a]	mg/m³[b]
Anisidine (o-, p-isomers) — Skin	0.1	0.5	—	—
** Antimony & Compounds (as Sb)	—	(0.5)	—	—
Antimony trioxide, handling and use (as Sb)	—	0.5	—	—
* Antimony trioxide production	—	A2	—	—
ANTU (α-Naphthyl thiourea)	—	0.3	—	0.9
Argon	F	—	F	F
* Arsenic & soluble compounds, as As	—	0.2	—	—
* Arsenic trioxide production	—	A2	—	—
Arsine	0.05	0.2	—	—
* Asbestos, see MINERAL DUSTS	—	A1a	—	A1a
Asphalt (petroleum) fumes	—	5	—	10
Atrazine	—	10	—	—
Azinphos-methyl — Skin	—	0.2	—	0.6
Barium (soluble compounds), as Ba	—	0.5	—	—
Baygon (propoxur)	—	0.5	—	2
* Baytex, see Fenthion	—	0.1	—	0.3
Benomyl	—	10	—	15
Benzene	10, A2	30, A2	25, A2	75, A2
** Benzidine production — Skin	—	(A1b)	—	(A1b)
p-Benzoquinone, see Quinone	0.1	0.4	0.3	2
Benzoyl peroxide	—	5	—	—
Benzo(a)pyrene	—	A2	—	A2
Benzyl chloride	1	5	—	—
Beryllium	—	0.002, A2	—	—
Biphenyl	0.2	1.5	0.6	4
Bismuth telluride	—	10	—	20
Bismuth telluride, Se-doped	—	5	—	10

Capital letters refer to Appendices.
Footnotes (a thru f) see Page 32.
*1980 Addition
**See Notice of Intended Changes.

Substance	ADOPTED VALUES		TENTATIVE VALUES	
	TWA		STEL	
	ppm[a]	mg/m^3[b]	ppm[a]	mg/m^3[b]
Borates, tetra, sodium salts, Anhydrous	—	1	—	—
Decahydrate	—	5	—	—
Pentahydrate	—	1	—	—
Boron oxide	—	10	—	20
Boron tribromide	1	10	3	30
C Boron trifluoride	1	3	—	—
Bromacil	1	10	2	20
Bromine	0.1	0.7	0.3	2
Bromine pentafluoride	0.1	0.7	0.3	2
Bromochloromethane, see Chlorobromomethane	200	1,050	250	1,300
Bromoform — Skin	0.5	5	—	—
Butadiene (1, 3-butadiene)	1,000	2,200	1,250	2,750
** Butane	(600)	(1,430)	(750)	(1,780)
Butanethiol, see Butyl mercaptan	0.5	1.5	—	—
2-Butanone, see Methyl ethyl ketone (MEK)	200	590	300	885
** 2-Butoxyethanol (Butyl Cellosolve) — Skin	(50)	(240)	(150)	(720)
n-Butyl acetate	150	710	200	950
sec-Butyl acetate	200	950	250	1,190
tert-Butyl acetate	200	950	250	1,190
Butyl acrylate	10	55	—	—
C n-Butyl alcohol — Skin	50	150	—	—
** sec-Butyl alcohol	(150)	(450)	—	—
tert-Butyl alcohol	100	300	150	450
C Butylamine — Skin	5	15	—	—
C tert-Butyl chromate (as CrO_3) — Skin	—	0.1	—	—
** n-Butyl glycidyl ether (BGE)	(50)	(270)	—	—
n-Butyl lactate	5	25	—	—
Butyl mercaptan	0.5	1.5	—	—
* o-sec-Butylphenol — Skin	5	30	—	—
p-tert-Butyltoluene	10	60	20	120

Capital letters refer to Appendices.
*1980 Addition.
**See Notice of Intended Changes.

Substance	ADOPTED VALUES		TENTATIVE VALUES	
	TWA		STEL	
	ppm[a]	mg/m^3[b]	ppm[a]	mg/m^3[b]
Cadmium, dust & salts (as Cd)	—	0.05	—	0.2
C Cadmium oxide fume (as Cd)	—	0.05	—	—
** Cadmium oxide production (as Cd)	—	(A2)	—	—
Calcium carbonate/ marble	—	E	—	20
Calcium cyanamide	—	0.5	—	1
Calcium hydroxide	—	5	—	—
Calcium oxide	—	2	—	—
Camphor, synthetic	2	12	3	18
Caprolactam Dust	—	1	—	3
Vapor	5	20	10	40
Captafol (Difolatan®) — Skin	—	0.1	—	—
Captan	—	5	—	15
Carbaryl (Sevin®)	—	5	—	10
Carbofuran (Furadan®)	—	0.1	—	—
Carbon black	—	3.5	—	7
Carbon dioxide	5,000	9,000	15,000	18,000
* Carbon disulfide — Skin	10	30	—	—
Carbon monoxide	50	55	400	440
Carbon tetrabromide	0.1	1.4	0.3	4
** Carbon tetrachloride — Skin	(10)	(65)	(20)	(130)
Carbonyl chloride (Phosgene)	0.1	0.4	—	—
** Carbonyl fluoride	(5)	(15)	—	—
Catechol (Pyrocatechol)	5	20	—	—
Cellulose (paper fiber)	—	E	—	20
Cesium hydroxide	—	2	—	—
Chlordane — Skin	—	0.5	—	2
Chlorinated camphene — Skin	—	0.5	—	1
Chlorinated diphenyl oxide	—	0.5	—	2
Chlorine	1	3	3	9
Chlorine dioxide	0.1	0.3	0.3	0.9
C Chlorine trifluoride	0.1	0.4	—	—

Capital letters refer to Appendices.
*1980 Addition.
**See Notice of Intended Changes.

Substance	ADOPTED VALUES		TENTATIVE VALUES	
	TWA		STEL	
	ppm[a]	mg/m^3[b]	ppm[a]	mg/m^3[b]
C Chloroacetaldehyde	1	3	—	—
α-Chloroacetophenone (Phenacyl chloride)	0.05	0.3	—	—
* Chloroacetyl chloride	0.05	0.2	—	—
Chlorobenzene (Monochlorobenzene)	75	350	—	—
** o-Chlorobenzylidene malononitrile — Skin	(0.05)	(0.4)	—	—
Chlorobromomethane	200	1,050	250	1,300
* 2-Chloro-1, 3-butadiene, see β Chloroprene — Skin	10	45	—	—
Chlorodifluoromethane	1,000	3,500	1,250	4,375
Chlorodiphenyl (42% Chlorine) — Skin	—	1	—	2
Chlorodiphenyl (54% Chlorine) — Skin	—	0.5	—	1
* 1-Chloro, 2, 3-epoxy-propane (Epichlorohydrin) — Skin	2	8	5	19
C 2-Chloroethanol (Ethylene chlorohydrin) — Skin	1	3	—	—
* Chloroethylene, see Vinyl chloride	5, A1a	10, A1a	—	—
Chloroform (Trichloromethane)	10, A2	50, A2	50	225
bis-Chloromethyl ether	0.001	A1a	—	A1a
** 1-Chloro-1-nitropropane	(20)	(100)	—	—
Chloropicrin	0.1	0.7	0.3	2
* β-Chloroprene — Skin	10	45	—	—
o-Chlorostyrene	50	285	75	430
o-Chlorotoluene — Skin	50	250	75	375
2-Chloro-6-(trichloromethyl) pyridine (N-Serve®)	—	10	—	20
Chlorpyrifos (Dursban®) — Skin	—	0.2	—	0.6
** Chromates, certain insoluble forms	—	(0.05,A1a)	—	(A1a)

Capital letters refer to Appendices.
Footnotes (a thru f) see Page 32.
*1980 Addition.
**See Notice of Intended Changes.

Substance	ADOPTED VALUES TWA ppm[a]	ADOPTED VALUES TWA mg/m^3[b]	TENTATIVE VALUES STEL ppm[a]	TENTATIVE VALUES STEL mg/m^3[b]
** Chromic acid and Chromates, (as Cr)	—	(0.05)	—	—
Chromite ore processing (chromate), as Cr	—	0.05, A1a	—	—
Chromium, Sol. chromic, chromous salts (as Cr)	—	0.5	—	—
Clopidol (Coyden®)	—	10	—	20
Coal tar pitch volatiles, as benzene solubles	—	0.2, A1a	—	A1a
** Cobalt metal, dust and fume (as Co)	—	(0.1)	—	—
Copper fume	—	0.2	—	—
Dusts & Mists (as Cu)	—	1	—	2
Corundum (Al_2O_3)	—	E	—	E
Cotton dust, raw	—	0.2[k]	—	0.6
Crag® herbicide, see Sodium 2, 4-dichlorophenoxyethyl sulfate	—	10	—	20
Cresol, all isomers — Skin	5	22	—	—
Crotonaldehyde	2	6	6	18
Crufomate®	—	5	—	20
Cumene — Skin	50	245	75	365
Cyanamide	—	2	—	—
Cyanides, as CN — Skin	—	5	—	—
Cyanogen	10	20	—	—
*C Cyanogen chloride	0.3	0.6	—	—
Cyclohexane	300	1,050	375	1,300
Cyclohexanol	50	200	—	—
** Cyclohexanone	(50)	(200)	—	—
Cyclohexene	300	1,015	—	—
Cyclohexylamine — Skin	10	40	—	—
Cyclonite — Skin	—	1.5	—	3
Cyclopentadiene	75	200	150	400
2, 4-D (2, 4-Dichlorophenoxy-acetic acid)	—	10	—	20
DDT (Dichlorodiphenyl-trichloroethane)	—	1	—	3

Capital letters refer to Appendices.
*1980 Addition.
**See Notice of Intended Changes.
k) See p. 35.

Substance	ADOPTED VALUES TWA ppm[a]	ADOPTED VALUES TWA mg/m^3[b]	TENTATIVE VALUES STEL ppm[a]	TENTATIVE VALUES STEL mg/m^3[b]
DDVP, see Dichlorvos — Skin	0.1	1	0.3	3
Decaborane — Skin	0.05	0.3	0.15	0.9
Demeton® — Skin	0.01	0.1	0.03	0.3
Diacetone alcohol (4-hydroxy-4-methyl-2-pentanone)	50	240	75	360
1, 2-Diaminoethane, see Ethylenediamine	10	25	—	—
Diazinon — Skin	—	0.1	—	0.3
Diazomethane	0.2	0.4	—	—
Diborane	0.1	0.1	—	—
** 1, 2-Dibromomethane, see Ethylene dibromide — Skin	(A1b)	(A1b)	—	—
Dibrom®	—	3	—	6
2-n-Dibutylaminoethanol — Skin	2	14	4	28
Dibutyl phosphate	1	5	2	10
Dibutyl phthalate	—	5	—	10
C Dichloracetylene	0.1	0.4	—	—
C o-Dichlorobenzene	50	300	—	—
p-Dichlorobenzene	75	450	110	675
3, 3'-Dichlorobenzidene — Skin	—	A2	—	A2
Dichlorodifluoromethane	1,000	4,950	1,250	6,200
1, 3-Dichloro-5, 5-dimethyl hydantoin	—	0.2	—	0.4
1, 1-Dichloroethane	200	810	250	1,010
* 1, 2-Dichloroethane, see Ethylene dichloride	10	40	15	60
1, 2-Dichloroethylene	200	790	250	1,000
Dichloroethyl ether — Skin	5	30	10	60
* Dichlorofluoromethane	10	40	—	—
** Dichloromethane, see Methylene chloride	(200)	(700)	(250)	(870)
C 1, 1-Dichloro-1-nitroethane	10	60	—	—
1, 2-Dichloropropane, see Propylene				

Capital letters refer to Appendices.
Footnotes (a thru f) see Page 32.
*1980 Addition.
**See Notice of Intented Changes.

Substance	ADOPTED VALUES TWA ppm[a]	ADOPTED VALUES TWA mg/m^3[b]	TENTATIVE VALUES STEL ppm[a]	TENTATIVE VALUES STEL mg/m^3[b]
dichloride	75	350	110	510
* Dichloropropene — Skin	1	5	10	50
* 2, 2-Dichloropropionic acid	1	6	—	—
Dichlorotetrafluoro-ethane	1,000	7,000	1,250	8,750
Dichlorvos (DDVP) — Skin	0.1	1	0.3	3
Dicrotophos (Bidrin®) — Skin	—	0.25	—	—
Dicyclopentadiene	5	30	—	—
Dicyclopentadienyl iron	—	10	—	20
Dieldrin — Skin	—	0.25	—	0.75
Diethylamine	3	15	—	—
Diethylaminoethanol — Skin	10	50	—	—
Diethylene triamine — Skin	1	4	—	—
Diethyl ether, see Ethyl ether	400	1,200	500	1,500
Diethyl phthalate	—	5	—	10
Difluorodibromomethane	100	860	150	1,290
**C Diglycidyl ether (DGE)	(0.5)	(3)	—	—
Dihydroxybenzene, see Hydroquinone	—	2	—	4
Diisobutyl ketone	25	150	—	—
Diisopropylamine — Skin	5	20	—	—
Dimethoxymethane, see Methylal	1,000	3,100	1,250	3,875
Dimethyl acetamide — Skin	10	35	15	50
Dimethylamine	10	18	—	—
Dimethylaminobenzene, see Xylidene — Skin	5	25	10	50
Dimethylaniline (N, N-Dimethylaniline) — Skin	5	25	10	50
Dimethylbenzene, see Xylene — Skin	100	435	150	650
Dimethyl carbamyl chloride	A2	A2	—	—

Capital letters refer to Appendices.
*1980 Addition.
**See Notice of Intended Changes.

Substance	ADOPTED VALUES TWA ppm[a)]	mg/m³[b)]	TENTATIVE VALUES STEL ppm[a)]	mg/m³[b)]
Dimethyl-1, 2-dibromo-2-dichloroethyl phosphate, see Dibrom	—	3	—	6
Dimethylformamide — Skin	10	30	20	60
2, 6-Dimethyl-4-heptanone, see Diisobutyl ketone	25	150	—	—
1, 1-Dimethylhydrazine — Skin	0.5	1	1	2
Dimethylphthalate	—	5	—	10
Dimethyl sulfate — Skin	0.1, A2	0.5, A2	—	—
Dinitrobenzene (all isomers) — Skin	0.15	1	0.5	3
Dinitro-o-cresol — Skin	—	0.2	—	0.6
3, 5-Dinitro-o-toluamide (Zoalene®)	—	5	—	10
Dinitrotoluene — Skin	—	1.5	—	5
** Dioxane, tech. grade — Skin	(50)	(180)	—	—
Dioxathion (Delnav®) — Skin	—	0.2	—	—
Diphenyl, see Biphenyl	0.2	1.5	0.6	4
Diphenylamine	—	10	—	20
C Diphenylmethane diisocyanate, see Methylene bisphenyl isocyanate (MDI)	0.02	0.2	—	—
Dipropylene glycol methyl ether — Skin	100	600	150	900
Diquat	—	0.5	—	1
Di-sec, octyl phthalate (Di-2-ethylhexyl-phthalate)	—	5	—	10
Disulfiram	—	2	—	5
Disulfoton (Disyston®)	—	0.1	—	0.3
2, 6-Ditert. butyl-p-cresol	—	10	—	20
Diuron	—	10	—	—
* Divinyl benzene	10	50	—	—
Dyfonate — Skin	—	0.1	—	—
Emery	—	E	—	20

Capital letters refer to Appendices.
*1980 Addition.
**See Notice of Intended Changes.

Substance	ADOPTED VALUES TWA ppm[a)]	mg/m³[b)]	TENTATIVE VALUES STEL ppm[a)]	mg/m³[b)]
Endosulfan (Thiodan®) — Skin	—	0.1	—	0.3
Endrin — Skin	—	0.1	—	0.3
* Epichlorohydrin — Skin	2	10	5	20
EPN — Skin	—	0.5	—	2
1, 2-Epoxypropane, see Propylene oxide	100	240	150	360
2, 3-Epoxy-1-propanol, see Glycidol	50	150	75	225
Ethane	F	—	F	—
Ethanethiol, see Ethyl mercaptan	0.5	1	2	3
Ethanolamine	3	8	6	15
Ethion (Nialate®) — Skin	—	0.4	—	—
** 2-Ethoxyethanol — Skin	(100)	(370)	(150)	(560)
** 2-Ethoxyethyl acetate (Cellosolve acetate) — Skin	(100)	(540)	(150)	(810)
Ethyl acetate	400	1,400	—	—
** Ethyl acrylate — Skin	(25)	(100)	—	—
Ethyl alcohol (Ethanol)	1,000	1,900	—	—
Ethylamine	10	18	—	—
Ethyl amyl ketone (5-Methyl-3-heptanone)	25	130	—	—
Ethyl benzene	100	435	125	545
Ethyl bromide	200	890	250	1,110
Ethylbutyl ketone (3-Heptanone)	50	230	75	345
Ethyl chloride	1,000	2,600	1,250	3,250
Ethylene	F	—	F	—
C Ethylene chlorohydrin — Skin	1	3	—	—
Ethylenediamine	10	25	—	—
** Ethylene dibromide	(A1b)	(A1b)	—	—
* Ethylene dichloride	10	40	15	60
Ethylene glycol, Particulate	—	10	—	20
** Vapor	(100)	(250)	(125)	(325)
**C Ethylene glycol dinitrate and/or Nitroglycerin				

Capital letters refer to Appendices.
*1980 Addition.
**See Notices of Intended Changes.
Footnotes (a thru f) see Page 32.

Substance	ADOPTED VALUES TWA ppm[a)]	mg/m³[b)]	TENTATIVE VALUES STEL ppm[a)]	mg/m³[b)]
— Skin	(0.2)	(2)	—	—
Ethylene glycol methyl ether acetate (Methyl Cellosolve® acetate) — Skin	25	120	35	170
** Ethylene oxide	(50)	(90)	(75)	(135)
Ethylenimine — Skin	0.5	1	—	—
Ethyl ether	400	1,200	500	1,500
Ethyl formate	100	300	150	450
Ethylidene chloride, see 1, 1-Dichloroethane	200	810	250	1,010
C Ethylidene norbornene	5	25	—	—
Ethyl mercaptan	0.5	1	2	3
** N-Ethylmorpholine — Skin	20	95	—	—
Ethyl silicate	10	85	30	255
Fensulfothion (Dasanit)	—	0.1	—	—
Fenthion	—	0.1	—	0.3
Ferbam	—	10	—	20
Ferrovanadium dust	—	1	—	0.3
Fluoride (as F)	—	2.5	—	—
Fluorine	1	2	2	4
** Fluorotrichloromethane, see Trichlorofluoromethane	(1,000)	(5,600)	(1,250)	(7,000)
C Formaldehyde	2	3	—	—
Formamide	20	30	30	45
Formic acid	5	9	—	—
** Furfural — Skin	(5)	(20)	(15)	(60)
** Furfuryl alcohol — Skin	(5)	(20)	(10)	(40)
** Gasoline	—	(B2)	—	(B2)
Germanium tetrahydride	0.2	0.6	0.6	1.8
Glass, fibrous[e)] or dust	—	10	—	—
*C Glutaraldehyde	0.2	0.7	—	—
Glycerin mist	—	E	—	E
** Glycidol (2, 3-Epoxy-1-propanol)	(50)	(150)	(75)	(225)
Glycol monoethyl ether, see 2-Ethoxyethanol — Skin	100	370	150	560
Graphite (Synthetic)	—	E	—	—
Guthion®, see				

Capital letters refer to Appendices.
*1980 Addition.
**See Notice of Intended Changes.

Substance	ADOPTED VALUES TWA ppm[a]	ADOPTED VALUES TWA mg/m³[b]	TENTATIVE VALUES STEL ppm[a]	TENTATIVE VALUES STEL mg/m³[b]
Azinphos-methyl — Skin	—	0.2	—	0.6
Gypsum	—	E	—	20
Hafnium	—	0.5	—	1.5
Helium	F	—	F	—
Heptachlor — Skin	—	0.5	—	2
Heptane (n-Heptane)	400	1,600	500	2,000
Hexachlorocyclopentadiene	0.01	0.1	0.03	0.3
** Hexachloroethane — Skin	(1)	(10)	(3)	(30)
Hexachloronaphthalene — Skin	—	0.2	—	0.6
Hexafluoroacetone	0.1	0.7	0.3	2
** Hexane (n-hexane)	(100)	(360)	(125)	(450)
Hexamethyl phosphoramide — Skin	A2	A2	—	—
2-Hexanone, see Methyl butyl ketone — Skin	25	100	40	165
** Hexone, see Methyl isobutyl ketone) — Skin	(100)	(410)	(125)	(510)
sec-Hexyl acetate	50	300	—	—
C Hexylene glycol	25	125	—	—
Hydrazine — Skin	0.1, A2	0.1, A2	—	—
Hydrogen	F	—	F	—
Hydrogenated terphenyls	0.5	5	—	—
Hydrogen bromide	3	10	—	—
C Hydrogen chloride	5	7	—	—
*C Hydrogen cyanide — Skin	10	10	—	—
Hydrogen fluoride (as F)	3	2.5	6	5
Hydrogen peroxide	1	1.5	2	3
Hydrogen selenide (as Se)	0.05	0.2	—	—
Hydrogen sulfide	10	14	15	21
Hydroquinone	—	2	—	4
* 2-Hydroxypropyl acrylate — Skin	0.5	3	—	—
Indene	10	45	15	70

Capital letters refer to Appendices.
*1980 Addition.
**See Notice of Intended Changes.

Substance	ADOPTED VALUES TWA ppm[a]	ADOPTED VALUES TWA mg/m³[b]	TENTATIVE VALUES STEL ppm[a]	TENTATIVE VALUES STEL mg/m³[b]
Indium & Compounds (as In)	—	0.1	—	0.3
C Iodine	0.1	1	—	—
Iodoform	0.6	10	1	20
Iron oxide fume (Fe_2O_3, as Fe)	B3	5	—	10
** Iron pentacarbonyl (as Fe)	(0.01)	(0.08)	—	—
Iron salts, soluble (as Fe)	—	1	—	2
Isoamyl acetate	100	525	125	655
Isoamyl alcohol	100	360	125	450
Isobutyl acetate	150	700	187	875
Isobutyl alcohol	50	150	75	225
C Isophorone	5	25	—	—
Isophorone diisocyanate — Skin	0.01	0.09	—	—
Isopropyl acetate	250	950	310	1,185
Isopropyl alcohol — Skin	400	980	500	1,225
Isopropylamine	5	12	10	24
* N-Isopropylaniline — Skin	2	10	5	20
Isopropyl ether	250	1,050	310	1,320
Isopropyl glycidyl ether (IGE)	50	240	75	360
Kaolin	—	E	—	20
Ketene	0.5	0.9	1.5	3
Lead, inorg., fumes & dusts (as Pb)	—	0.15	—	0.45
** Lead arsenate (as Pb)	—	(0.15)	—	(0.45)
Lead chromate (as Cr)	—	0.05, A2	—	—
Limestone	—	E	—	20
Lindane — Skin	—	0.5	—	1.5
Lithium hydride	—	0.025	—	—
L.P.G. (Liquified petroleum gas)	1,000	1,800	1,250	2,250
Magnesite	—	E	—	20
Magnesium oxide fume (as Mg)	—	10	—	—
Malathion — Skin	—	10	—	—
Maleic anhydride	0.25	1	—	—

Capital letters refer to Appendices.
*1980 Addition.
**See Notice of Intended Changes.

Substance	ADOPTED VALUES TWA ppm[a]	ADOPTED VALUES TWA mg/m³[b]	TENTATIVE VALUES STEL ppm[a]	TENTATIVE VALUES STEL mg/m³[b]
C Manganese & Compounds (as Mn)	—	5	—	—
Manganese fume (as Mn)	—	1	—	3
Manganese cyclopentadienyl tricarbonyl (as Mn) — Skin	—	0.1	—	0.3
Manganese Tetroxide	—	1	—	—
Marble/calcium carbonate	—	E	—	20
Mercury (Alkyl compounds) — Skin, (as Hg)	—	0.01	—	0.03
** Mercury (All forms except alkyl), as Hg	—	(0.05)	—	(0.15)
** Mesityl oxide	(25)	(100)	—	—
Methane	F	—	F	—
Methanethiol, see Methyl mercaptan	0.5	1	—	—
Methomyl (Lannate®) — Skin	—	2.5	—	—
Methoxychlor	—	10	—	—
2-Methoxyethanol — Skin (Methyl Cellosolve®)	25	80	35	120
Methyl acetate	200	610	250	760
Methyl acetylene (propyne)	1,000	1,650	1,250	2,040
Methyl acetylene-propadiene mixture (MAPP)	1,000	1,800	1,250	2,250
Methyl acrylate — Skin	10	35	—	—
Methylacrylonitrile — Skin	1	3	2	6
Methylal (dimethoxymethane)	1,000	3,100	1,250	3,875
Methyl alcohol (methanol) — Skin	200	260	250	310
Methylamine	10	12	—	—
Methyl amyl alcohol, see Methyl isobutyl carbinol — Skin	25	100	40	160

Capital letters refer to Appendices.
**See Notice of Intended Changes.

Substance	ADOPTED VALUES TWA ppm[a]	ADOPTED VALUES TWA mg/m^3[b]	TENTATIVE VALUES STEL ppm[a]	TENTATIVE VALUES STEL mg/m^3[b]
** Methyl n-amyl ketone (2-Heptanone)	(100)	(465)	(150)	(710)
** Methyl bromide — Skin	(15)	(60)	—	—
** Methyl butyl ketone — Skin	(25	(100)	(40)	(165)
Methyl Cellosolve® — Skin see 2-Methoxyethanol	25	80	35	120
Methyl Cellosolve® acetate — Skin, see Ethylene glycol monomethyl ether acetate	25	120	35	170
** Methyl chloride	(100)	(210)	(125)	(260)
Methyl chloroform (1, 1, 1-Trichloroethane)	350	1,900	450	2,450
Methyl 2-cyanoacrylate	2	8	4	16
Methylcyclohexane	400	1,600	500	2,000
Methylcyclohexanol	50	235	75	350
o-Methycyclohexanone — Skin	50	230	75	345
Methylcyclopentadienyl manganese tricarbonyl (as MN) — Skin	—	0.2	—	0.6
Methyl demeton — Skin	—	0.5	—	1.5
C Methylene bisphenyl isocyanate (MDI)	0.02	0.2	—	—
** Methylene chloride (dichloromethane)	(200)	(700)	(250)	(870)
4, 4′-Methylene bis (2-chloroaniline) — Skin	0.02, A2	0.22, A2	—	—
C Methylene bis (4-cyclo-hexylisocyanate)	0.01	0.11	—	—
* 4, 4-Methylene dianiline	0.1	0.8	0.5	4
Methyl ethyl ketone (MEK)	200	590	300	885
C Methyl ethyl ketone peroxide	0.2	1.5	—	—
Methyl formate	100	250	150	375
C Methyl hydrazine — Skin	0.2	0.35	—	—
** Methyl iodide — Skin	(5)	(28)	(10)	(56)

Capital letters refer to Appendices.
*1980 Addition.
**See Notice of Intended Changes.

Substance	ADOPTED VALUES TWA ppm[a]	ADOPTED VALUES TWA mg/m^3[b]	TENTATIVE VALUES STEL ppm[a]	TENTATIVE VALUES STEL mg/m^3[b]
** Methyl isoamyl ketone	(100)	(475)	(150)	(710)
Methyl isobutyl carbinol — Skin	25	100	40	165
Methyl isobutyl ketone	100	410	125	510
Methyl isocyanate — Skin	0.02	0.05	—	—
Methyl mercaptan	0.5	1	—	—
Methyl methacrylate	100	410	125	510
Methyl parathion — Skin	—	0.2	—	0.6
Methyl propyl ketone	200	700	250	875
**C Methyl silicate	(5)	(30)	—	—
**C α-Methyl styrene	(100)	(480)	—	—
Molybdenum (as Mo)				
Soluble compounds	—	5	—	10
Insoluble compounds	—	10	—	20
Monocrotophos (Azodrin®)	—	0.25	—	—
** Monomethyl aniline — Skin	(2)	(9)	(4)	(18)
Morpholine — Skin	20	70	30	105
Naphthalene	10	50	15	75
β-Naphthylamine	—	A1b	—	A1b
Neon	F	—	F	—
Nickel carbonyl (as Ni)	0.05	0.35	—	—
Nickel metal	—	1	—	—
Nickel, soluble compounds (as Ni)	—	0.1	—	0.3
Nickel sulfide roasting, fume & dust (as Ni)	—	1, A1a	—	—
Nicotine — Skin	—	0.5	—	1.5
Nitric acid	2	5	4	10
Nitric oxide	25	30	35	45
p-Nitroaniline — Skin	1	6	2	12
Nitrobenzene — Skin	1	5	2	10
p-Nitrochlorobenzene — Skin	—	1	—	2
4-Nitrodiphenyl	—	A1b	—	A1b
Nitroethane	100	310	150	465
**C Nitrogen dioxide	(5)	(9)	—	—
Nitrogen trifluoride	10	30	15	45
**C Nitroglycerin — Skin	(0.2)	(2)	—	—

Capital letters refer to Appendices.
Footnotes (a thru f) see Page 32.
*1980 Addition.
**See Notice of Intended Changes.

Substance	ADOPTED VALUES TWA ppm[a]	ADOPTED VALUES TWA mg/m^3[b]	TENTATIVE VALUES STEL ppm[a]	TENTATIVE VALUES STEL mg/m^3[b]
Nitromethane	100	250	150	375
** 1-Nitropropane	(25)	(90)	(35)	(135)
*C 2-Nitropropane	25, A2	90, A2	—	—
N-Nitrosodimethylamine (dimethylnitrosoamine) — Skin	—	A2	—	A2
** Nitrotoluene — Skin	(5)	(30)	(10)	(60)
Nitrotrichloromethane, see Chloropicrin	0.1	0.7	0.3	2
Nonane	200	1,050	250	1,300
Octachloronaphthalene — Skin	—	0.1	—	0.3
Octane	300	1,450	375	1,800
Oil mist, mineral	—	5[e]	—	10
Osmium tetroxide (as Os)	0.0002	0.002	0.0006	0.006
Oxalic acid	—	1	—	2
Oxygen difluoride	0.05	0.1	0.15	0.3
Ozone	0.1	0.2	0.3	0.6
Paraffin wax fume	—	2	—	6
Paraquat, respirable sizes	—	0.1	—	—
Parathion — Skin	—	0.1	—	0.3
Particulate polycyclic aromatic hydro-carbons (PPAH), see Coal tar pitch volatiles	—	0.2, A1a	—	A1a
Pentaborane	0.005	0.01	0.015	0.03
Pentachloronaphthalene	—	0.5	—	2
Pentachlorophenol — Skin	—	0.5	—	1.5
Pentaerythritol	—	E	—	20
Pentane	600	1,800	750	2,250
2-Pentanone, see Methyl propyl ketone	200	700	250	875
** Perchloroethylene — Skin	(100)	(670)	(150)	(1,000)
Perchloromethyl mercaptan	0.1	0.8	—	—

Capital letters refer to Appendices.
Footnotes (a thru f) see Page 32.
*1980 Addition.
**See Notice of Intended Changes.

Substance	ADOPTED VALUES TWA ppm[a)]	mg/m³[b)]	TENTATIVE VALUES STEL ppm[a)]	mg/m³[b)]
Perchloryl fluoride	3	14	6	28
Phenol — Skin	5	19	10	38
Phenothiazine — Skin	—	5	—	10
* N-Phenyl-beta-naphthylamine	A2	A2	—	—
p-Phenylene diamine — Skin	—	0.1	—	—
Phenyl ether (vapor)	1	7	2	14
** Phenyl ether-Diphenyl mixture (vapor)	(1)	(7)	(2)	(14)
Phenylethylene, see Styrene, monomer	100	420	125	525
** Phenyl glycidyl ether (PGE)	(10)	(60)	(15)	(90)
Phenylhydrazine — Skin	5	20	10	45
Phenyl mercaptan	0.5	2	—	—
C Phenylphosphine	0.05	0.25	—	—
Phorate (Thimet®) — Skin	—	0.05	—	0.2
Phosdrin (Mevinphos®) — Skin	0.01	0.1	0.03	0.3
Phosgene (carbonyl chloride)	0.1	0.4	—	—
Phosphine	0.3	0.4	1	1
Phosphoric acid	—	1	—	3
Phosphorus (yellow)	—	0.1	—	0.3
* Phosphorus pentachloride	0.1	1	—	—
Phosphorus pentasulfide	—	1	—	3
** Phosphorus trichloride	(0.5)	(3)	—	—
Phthalic anhydride	1	6	4	24
m-Phthalodinitrile	—	5	—	—
Picloram (Tordon®)	—	10	—	20
Picric acid — Skin	—	0.1	—	0.3
Pival® (2-Pivalyl-1, 3-indandione)	—	0.1	—	0.3
Plaster of Paris	—	E	—	20
Platinum (Soluble salts) as Pt	—	0.002	—	—
Polychlorobiphenyls, see Chlorodiphenyls — Skin	—	—	—	—

Capital letters refer to Appendices.
*1980 Addition.
**See Notice of Intended Changes.

Substance	ADOPTED VALUES TWA ppm[a)]	mg/m³[b)]	TENTATIVE VALUES STEL ppm[a)]	mg/m³[b)]
Polytetrafluoroethylene decomposition products	—	B1	—	B1
C Potassium hydroxide	—	2	—	—
Propane	F	—	F	—
Propargyl alcohol — Skin	1	2	3	6
** β-Propiolactone	—	(A2)	—	(A2)
* Propionic acid	10	30	15	45
n-Propyl acetate	200	840	250	1,050
Propyl alcohol — Skin	200	500	250	625
n-Propyl nitrate	25	105	40	470
Propylene	F	—	F	—
Propylene dichloride (1, 2-Dichloropropane)	75	350	110	510
**C Propylene glycol dinitrate — Skin	(0.2)	(2)	—	—
Propylene glycol monomethyl ether	100	360	150	540
** Propylene imine — Skin	(2)	(5)	—	—
** Propylene oxide	(100)	(240)	(150)	(360)
Propyne, see Methyl acetylene	1,000	1,650	1,250	2,040
Pyrethrum	—	5	—	10
Pyridine	5	15	10	30
Quinone	0.1	0.4	0.3	2
RDX, see Cyclonite — Skin	—	1.5	—	3
Resorcinol	10	45	20	90
** Rhodium, Metal fume and dusts (as Rh)	—	(0.1)	—	(0.3)
Soluble salts (as Rh)	—	0.001	—	0.003
Ronnel	—	10	—	—
Rosin core solder pyrolysis products (as formaldehyde)	—	0.1	—	0.3
Rotenone (commercial)	—	5	—	10
Rouge	—	E	—	20
Rubber solvent (Naphtha)	400	1,600	—	—
Selenium compounds (as Se)	—	0.2	—	—

Capital letters refer to Appendices.
*1980 Addition.
**See Notice of Intended Changes.

Substance	ADOPTED VALUES TWA ppm[a)]	mg/m³[b)]	TENTATIVE VALUES STEL ppm[a)]	mg/m³[b)]
Selenium hexafluoride, (as Se)	0.05	0.2	—	—
Sevin® (see Carbaryl)	—	5	—	10
** Silane (see Silicon tetrahydride)	(0.5)	(0.7)	(1)	(1.5)
Silicon	—	E	—	20
Silicon carbide	—	E	—	20
** Silicon tetrahydride (Silane)	(0.5)	(0.7)	(1)	(1.5)
* Silver, metal	—	0.1	—	—
** Silver, soluble compounds, as Ag	—	(0.01)	—	(0.03)
C Sodium azide	0.1	0.3	—	—
* Sodium bisulfite	—	5	—	—
Sodium 2, 3-dichloro-phenoxyethyl sulfate	—	10	—	20
Sodium fluoroacetate (1080) — Skin	—	0.05	—	0.15
C Sodium hydroxide	—	2	—	—
* Sodium metabisulfite	—	5	—	—
Starch	—	E	—	20
Stibine	0.1	0.5	0.3	1.5
** Stoddard solvent	100	(575)	(125)	(720)
Strychnine	—	0.15	—	0.45
** Styrene, monomer (Phenylethylene)	(100)	(420)	(125)	(525)
C Subtilisins (Proteolytic enzymes as 100% pure crystalline enzyme)	—	0.00006[m)]	—	—
Sucrose	—	E	—	20
* Sulfur dioxide	2	5	5	10
Sulfur hexafluoride	1,000	6,000	1,250	7,500
Sulfuric acid	—	1	—	—
Sulfur monochloride	1	6	3	18
Sulfur pentafluoride	0.025	0.25	0.075	0.75
Sulfur tetrafluoride	0.1	0.4	0.3	1
Sulfuryl fluoride	5	20	10	40
Systox, see Demeton® — Skin	0.01	0.1	0.03	0.3
2, 4, 5-T	—	10	—	20

Capital letters refer to Appendices.
*1980 Addition.
**See Notice of Intended Changes.
m) See Page 35.

Substance	ADOPTED VALUES TWA ppm[a]	mg/m³[b]	TENTATIVE VALUES STEL ppm[a]	mg/m³[b]
Tantalum	—	5	—	10
TEDP — Skin	—	0.2	—	0.6
Teflon® decomposition products	—	B1	—	B1
Tellurium & compounds (as Te)	—	0.1	—	—
Tellurium hexafluoride, as Te	0.02	0.2	—	—
TEPP — Skin	0.004	0.05	0.01	0.2
*C Terphenyls	0.5	5	—	—
1, 1, 1, 2-Tetrachloro-2, 2-difluoroethane	500	4,170	625	5,210
1, 1, 2, 2-Tetrachloro-1, 2-difluoroethane	500	4,170	625	5,210
** 1, 1, 2, 2-Tetrachloroethane — Skin	(5)	(35)	(10)	(70)
Tetrachloroethylene, see Perchloroethylene — Skin	100	670	150	1,000
Tetrachloromethane, see Carbon tetrachloride — Skin	10	65	20	130
Tetrachloronaphthalene	—	2	—	4
Tetraethyl lead (as Pb) — Skin	—	0.100[f]	—	0.3
Tetrahydrofuran	200	590	250	735
Tetramethyl lead (as Pb) — Skin	—	0.150[f]	—	0.5
Tetramethyl succinonitrile — Skin	0.5	3	2	9
* Tetrasodium pyrophosphate	—	1	—	—
Tetranitromethane	1	8	—	—
Tetryl (2, 4, 6-trinitrophenyl-methylnitramine) — Skin	—	1.5	—	3.0
Thallium, soluble compounds (as Tl) — Skin	—	0.1	—	—

Capital letters refer to Appendices.
Footnotes (a thru f) see Page 32.
*1980 Addition.
**See Notice of Intended Changes.

Substance	ADOPTED VALUES TWA ppm[a]	mg/m³[b]	TENTATIVE VALUES STEL ppm[a]	mg/m³[b]
4, 4'-Thiobis (6-tert. butyl-m-cresol)	—	10	—	20
Thioglycolic acid	1	5	—	—
Thiram®	—	5	—	10
** Tin, inorganic compounds, except SnH_4 and SnO_2 (as Sn)	—	(2)	—	(4)
Tin, organic compounds (as Sn) — Skin	—	0.1	—	0.2
** Tin oxide (as Sn)	—	(E)	—	(20)
Titanium dioxide (as Ti)	—	E	—	20
Toluene (toluol) — Skin	100	375	150	560
**C Toluene-2, 4-diisocyanate (TDI)	(0.02)	(0.14)	—	—
** o-Toluidine	(5)	(22)	(10)	(44)
Toxaphene, see Chlorinated camphene — Skin	—	0.5	—	2.0
** Tributyl phosphate	—	(5)	—	(5)
* Trichloroacetic acid	—	1	—	—
1, 2, 4-Trichlorobenzene	5	40	—	—
1, 1, 1-Trichloroethane, see Methyl chloroform	350	1,900	450	2,380
1, 1, 2-Trichloroethane — Skin	10	45	20	90
** Trichloroethylene	(100)	(535)	(150)	(800)
** Trichlorofluoromethane	(1,000)	(5,600)	(1,250)	(7,000)
Trichloromethane, see Chloroform	10, A2	50, A2	—	—
Trichloronaphthalene	—	5	—	10
1, 2, 3-Trichloropropane	50	300	75	450
1, 1, 2-Trichloro 1, 2, 2-trifluoroethane	1,000	7,600	1,250	9,500
Tricyclohexyltin hydroxide (Plictran®)	—	5	—	10
Triethylamine	25	100	40	160
Trifluorobromomethane	1,000	6,100	1,200	7,300
Trimethyl benzene	25	125	35	170
** Trimethyl phosphite	(0.5)	(2.6)	—	—
2, 4, 6-Trinitrophenol, see Picric acid — Skin	—	0.1	—	0.3

Capital letters refer to Appendices.
*1980 Addition.
**See Notice of Intended Changes.

Substance	ADOPTED VALUES TWA ppm[a]	mg/m³[b]	TENTATIVE VALUES STEL ppm[a]	mg/m³[b]
2, 4, 6-Trinitrophenyl-methylnitramine, see Tetryl — Skin	—	1.5	—	3.0
**C 2, 4, 6-Trinitrotoluene (TNT)	—	(0.5)	—	—
Triorthocresyl phosphate	—	0.1	—	0.3
Triphenyl amine	—	5	—	—
Triphenyl phosphate	—	3	—	6
Tungsten & compounds, as W				
Soluble	—	1	—	3
Insoluble	—	5	—	10
Turpentine	100	560	150	840
Uranium (natural) soluble & insoluble compounds, as U	—	0.2	—	0.6
** Vanadium (V_2O_5), as V				
Dust	—	(0.5)	—	(1.5)
C Fume	—	(0.05)	—	—
Valeraldehyde	50	175	—	—
Vinyl acetate	10	30	20	60
** Vinyl benzene, see Styrene	(100)	(420)	(125)	(525)
* Vinyl bromide	5, A2	20, A2	—	—
* Vinyl chloride	5, A1a	10, A1a	—	—
** Vinyl cyanide, see Acrylonitrile	(A1b)	(A1b)	—	—
Vinyl cyclohexene dioxide	10	60	—	—
Vinylidene chloride	10	40	20	80
** Vinyl toluene	(100)	(480)	(150)	(720)
VM & P Naphtha	300	1,350	400	1,800
Warfarin	—	0.1	—	0.3
Welding fumes (NOC)†	—	5, B3	—	B3
** Wood dust (nonallergenic)	—	(5)	—	(10)
Xylene (o-, m-, p-isomers) — Skin	100	435	150	655
C m-Xylene α, α'-diamine	—	0.1	—	—

*1980 Addition
**See Notice of Intended Changes.
Capital letters refer to Appendices.
†(NOC) not otherwise classified

Substance	ADOPTED VALUES TWA ppm[a]	mg/m³[b]	TENTATIVE VALUES STEL ppm[a]	mg/m³[b]
** Xylidene — Skin	(5)	(25)	(10)	(50)
Yttrium	—	1	—	3
Zinc chloride fume	—	1	—	2
Zinc chromate (as Cr)	—	0.05, A2	—	—
Zinc oxide fume	—	5	—	10
Zinc stearate	—	E	—	20
Zirconium compounds (as Zr)	—	5	—	10

**See notice of intended changes

a) Parts of vapor or gas per million parts of contaminated air by volume at 25°C and 760 mm. Hg. pressure.
b) Approximate milligrams of substance per cubic meter of air.
d) < 7 μm in diameter.
e) As sampled by method that does not collect vapor.
f) For control of general room air, biologic monitoring is essential for personnel control.

Radioactivity: The Committee accepts the philosophy and recommendations of the National Council on Radiation Protection and Measurements (NCRP) for the ionizing radiation TLV. The NCRP is charted by Congress to, in part, collect, analyze, develop and disseminate information and recommendations about protection against radiation and about radiation measurements, quantities and units, including development of basic concepts in these areas. NCRP Report No. 39 (reference 1) provides basic philosophy and concepts leading to protection criteria established in the same report. Other NCRP reports address specific areas of radiation protection and, collectively, provide an excellent basis for establishing a sound program for radiation control. The Committee recommends the listed references as substantative documentation of a sound basis for ionizing radiation protection. The Committee also strongly recommends that all exposures to ionizing radiations be kept low as reasonably achievable within the stated guidance.

References:
1. "Basic Radiation Protection Criteria," NCRP Report No. 39, issued January 15, 1971.
2. Maximum Permissible Body Burdens and Maximum Permissible Concentrations of Radionuclides in Air and in Water for Occupational Exposure," US Department of Commerce, National Bureau of Standards Handbook 69, issued June 5, 1959, with Addendum 1 issued August 1963. Available as NCRP Report No. 22.

The above documents, as well as information on numerous other NCRP Reports addressing specific subjects in ionizing radiation protection are available from: NCRP Publications, PO Box 30175, Washington, DC 20014.

MINERAL DUSTS

Substance

SILICA, SiO₂

Crystalline

Quartz — TLV in mppcf[g]:

$$\frac{300^{h)}}{\%\ \text{quartz} + 10}$$

TLV for respirable dust in mg/m³:

$$\frac{10\ \text{mg/m}^{3\,i)}}{\%\ \text{Respirable quartz} + 2}$$

TLV for "total dust," respirable and nonrespirable:

$$\frac{30\ \text{mg/m}^3}{\%\ \text{quartz} + 3}$$

Cristobalite .... Use one-half the value calculated from the count or mass formulae for quartz.

Tridymite — Use one-half the value calculated from formulae for quartz.

Silica, fused — Use quartz formulae.

Tripoli — Use respirable[n] mass quartz formula

***Amorphous* .... (20 mppcf[g])

**See Notice of Intended Changes.
g), h), i) See p. 34.
n) See p. 35.

SILICATES (< 1% quartz)

*Asbestos	
Amosite	0.5 fiber > 5μm/cc, A1a
Chrysotile	2 fibers > 5μm/cc, A1a
Crocidolite	0.2 fiber > 5μm/cc, A1a
Other forms	2 fibers > 5μm/cc, A1a
Mica	20 mppcf
Mineral wool fiber	10 mg/m³
Perlite	30 mppcf
Portland Cement	30 mppcf
Soapstone	20 mppcf
**Talc (nonasbestiform)	(20 mppcf)
**Talc (fibrous), (use Asbestos limit.)	

COAL DUST

2 mg/m³ (respirable dust fraction < 5% quartz).
If > 5% quartz, use respirable mass formula.

NUISANCE PARTICULATES

(see Appendix E)

30 mppcf or 10 mg/m³[i]

of total dust < 1% quartz, or, 5 mg/m³ respirable dust.

Conversion factors:
mppcf × 35.3 = Million particles per cubic meter
= particles per cc

g) Millions of particles per cubic foot of air, based on impinger samples counted by light-field technics.
h) The percentage of quartz in the formula is the amount determined from airborne samples, except in those instances in which other methods have been shown to be applicable.
i) Both concentration and percent quartz for the application of this limit are to be determined from the fraction passing a size-selector with the following characteristics:

Aerodynamic Diameter (μm) (unit density sphere)	% passing selector
≤ 2	90
2.5	75
3.5	50
5.0	25
10	0

*1980 Addition.
**See Notice of Intended Changes.

j) containing < 1% quartz; if quartz content > 1%, use formulae for quartz.

k) Lint-free dust as measured by the vertical-elutriator, cotton-dust sampler described in the Transactions of the National Conference on Cotton Dust, J. R. Lynch, pg. 33, May 2, 1970.

l) As determined by the membrane filter method at 400–450X magnification (4 mm objective) phase contrast illumination.

m) Based on "high volume" sampling.

n) "Respirable" dust as defined by the British Medical Research Council Criteria (1) and as sampled by a device producing equivalent results (2).

(1) Hatch, T. E. and Gross, P., Pulmonary Deposition and Retention of Inhaled Aerosols, p. 149. Academic Press, New York, New York, 1964.

(2) Interim Guide for Respirable Mass Sampling, AIHA Aerosol Technology Committee, AHIA J. *31:* 2, 1970, p. 133.

NOTICE OF INTENDED CHANGES (for 1980)

These substances, with their corresponding values, comprise those for which either a limit has been proposed for the first time, or for which a change in the "Adopted" listing has been proposed. In both cases, the proposed limits should be considered trial limits that will remain in the listing for a period of at least two years. If, after two years no evidence comes to light that questions the appropriateness of the values herein, the values will be reconsidered for the "Adopted" list. Documentation is available for each of these substances.

Substance	TWA ppm[a]	TWA mg/m³[b]	STEL ppm[a]	STEL mg/m³[b]
Acetone	750	1780	1000	2375
Acrylic acid	10	30	—	—
† Acrylonitrile	2, A1a	4.5, A1a	—	—
† Benzidine — Skin	—	A1b	—	—
Butane	800	1900	—	—
2-Buoxyethanol (Butyl Cellosolve®) — Skin	25	120	75	360
sec-Butyl alcohol	100	305	150	455
n-Butyl glycidyl ether (BGE)	25	135	—	—
† Cadmium oxide production	—	0.05	—	—
Carbon tetrachloride — Skin	5, A2	30, A2	20, A2	125, A2
Carbonyl fluoride	2	5	5	15
†C o-Chlorobenzylidene malononitrile	0.05	0.4	—	—
† Chloromethyl methyl ether	A2	A2	—	—
1-Chloro-1-nitropropane	2	10	—	—
Chloropentafluoroethane	1000	6320	—	—
Chromium metal	—	0.5	—	—
Chromium (II) compounds, as Cr	—	0.5	—	—
Chromium (III) compounds, as Cr	—	0.5	—	—
Chromium (VI) compounds, as Cr				
Water soluble Cr VI compounds	—	0.05	—	—
Certain water insoluble Cr VI compounds	—	0.05, A1a	—	—
† Chromyl chloride	0.025	0.15	—	—
Chrysene	A2	A2	—	—
Cobalt metal, dust & fume, as Co	—	0.05	—	0.1
Cyclohexanone	25	100	100	400
Cyclopentane	600	1720	900	2580
1, 2-Dibromoethane, see Ethylene dibromide — Skin	A2	A2	—	—
1, 1-Dichloro-1-nitroethane	2	10	10	60
Diethylamine	10	30	25	75
Diethyl ketone	200	705	--	—
Diglycidyl ether (DGE)	0.1	0.5	—	—

Capital letters refer to Appendices.
†1980 Revision or Addition.

Substance	TWA ppm[a]	TWA mg/m³[b]	STEL ppm[a]	STEL mg/m³[b]
Dioxane, tech. grade — Skin	25	90	100	360
Dipropyl ketone	50	235	—	—
† Enfluorane	75	575	—	—
2-Ethoxyethanol — Skin	50	185	100	370
2-Ethoxyethyl acetate (Cellosolve® acetate) — Skin	50	270	100	540
Ethyl acrylate — Skin	5	20	25	100
† Ethylene dibromide — Skin	A2	A2	—	—
C Ethylene glycol, vapor	50	125	—	—
Ethylene glycol dinitrate	0.02	0.2	0.04	0.4
Ethylene oxide	10	20	—	—
† N-Ethylmorpholine — Skin	5	23	20	95
Furfural — Skin	2	8	10	40
† Furfuryl alcohol — - Skin	10	40	15	60
† Gasoline	300	900	500	1500
Glycidol (2, 3-Epoxy-1-propanol)	25	75	100	300
† Halothane	50	400	—	—
† Hexachlorobutadiene	0.02, A2	0.24, A2	—	—
† Hexachloroethane — Skin	10	100	—	—
† Hexane (n-Hexane)	50	180	—	—
(other Isomers)	500	1800	1,000	3,600
Hexone, see Methyl isobutyl ketone — Skin	50	205	75	300
† Iron pentacarbonyl, as Fe	0.1	0.8	0.2	0.16
† Isooctyl alcohol	50	270	—	—
Isopropoxyethanol	25	105	75	320
† Lead arsenate, as $Pb_3(AsO_4)_2$	—	0.15	—	0.45
† Mercury (All forms except alkyl) — Skin, as Hg				
Vapor	—	0.05	—	—
Aryl & inorganic compounds	—	0.1	—	—
Mesityl oxide	15	60	25	100
Methacrylic acid	20	70	—	—
† 4-Methoxyphenol	—	5	—	—
Methyl n-amyl ketone (2-Heptanone)	50	235	100	465
† N-Methyl aniline — Skin	0.5	2	1	5

Capital letters refer to Appendices.
†1980 Revision or Addition.

Substance	TWA ppm[a]	TWA mg/m³[b]	STEL ppm[a]	STEL mg/m³[b]
Methyl bromide — Skin	5	20	15	60
Methyl n-butyl ketone	5	20	—	—
Methyl chloride	50	105	100	205
Methylene chloride (dichloromethane)	100	360	500	1,700
Methyl iodide — Skin	2, A2	10, A2	5, A2	30, A2
† Methyl isoamyl ketone	50	240	—	—
Methyl isobutyl ketone — Skin	50	205	75	300
Methyl isopropyl ketone	200	705	—	—
Methyl silicate	1	6	5	30
α-Methyl styrene	50	240	100	485
† p-Nitroaniline — Skin	—	3	—	—
† p-Nitrochlorobenzene — Skin	0.5	3	—	—
Nitrogen dioxide	3	6	5	10
Nitroglycerin	0.02	0.2	0.04	0.4
1-Nitropropane	15	55	25	90
† Nitrotoluene — Skin	2	11	—	—
† Perchloroethylene — Skin	50	335	—	—
† Persulfates, alkali metal, as S_2O_8	—	2	—	—
Phenyl ether — Diphenyl mixture (vapor)	0.5	4	2	16
† Phenyl glycidyl ether (PGE)	1	6	—	—
† Phosphorus oxychloride	0.1	0.6	0.5	3
† Phosphorus trichloride	0.2	1.5	0.5	3
† Piperazine dihydrochloride	—	5	—	—
Platinum metal	—	1	—	—
β-Propiolactone	0.5, A2	1.5, A2	1, A2	3, A2
Propylene glycol dinitrate	0.02	0.2	0.04	0.4
† Propylene imine — Skin	2, A2	5, A2	—	—
Propylene oxide	20	50	—	—
† Rhodium, Metal	—	1	—	—
Insoluble compounds, as Rh	—	0.1	—	0.3
† Silicon tetrahydride (Silane)	5	7	—	—
Silver, soluble compounds, as Ag	—	0.01	—	—
† Stoddard solvent	100	525	200	1,050
Styrene, monomer (phenylethylene)	50	215	100	425

Capital letters refer to Appendices.
†1980 Revision or Addition.

Substance	TWA ppm[a]	TWA mg/m³[b]	STEL ppm[a]	STEL mg/m³[b]
† 1, 1, 2, 2-Tetrachloroethane — Skin	1	7	5	35
† Tin, tin oxide & inorganic compounds, except SnH_4, as Sn	—	2	—	4
† o-Tolidine	A2	A2	—	—
Toluene-2, 4-diisocyanate (TDI)	0.005	0.04	0.02	0.15
† o-Toluidine	2	9	—	—
Tributyl phosphate	0.2	2.5	0.4	5
Trichloroethylene	50	270	150	805
†C Trichlorofluoromethane	1,000	5,600	—	—
Trimellitic anhydride	0.005	0.04	—	—
† Trimethyl phosphite	2	10	5	25
† 2, 4, 6-Trinitrotoluene (TNT) — Skin	—	0.5	—	3
† Vanadium, as V_2O_5, dust & fume	—	0.05	—	—
Vinyl toluene	50	240	100	485
Wood dust, hard wood (as in furniture making)	—	1	—	—
† Xylidine — Skin	2	10	—	—

NOTICE OF INTENDED CHANGES MINERAL DUSTS

Substance	TLV
Diatomaceous earth, natural	1.5 mg/m³, Respirable dust
Silica, amorphous	6 mg/m³, Total dust (all sampled sizes) 3 mg/m³, Respirable dust (< 5 μm)
† Talc (containing no fibers)	15 mppcf or 2 mg/m³, Respirable Dust
† Talc (fiber-containing)	2 fibers/cc, > 5 μm in length

Capital letters refer to Appendices.
†1980 Revision or Addition.

APPENDIX A CARCINOGENS

The Committee lists below those substances in industrial use that have proven carcinogenic in man, or have induced cancer in animals under appropriate experimental conditions. Present listing of those substances carcinogenic for man takes two forms: Those for which a TLV has been assigned (1a) and those for which environmental conditions have not been sufficiently defined to assign a TLV (1b).

A1a. *Human Carcinogens*. Substances, or substances associated with industrial processes, recognized to have carcinogenic or cocarcinogenic potential, with an assigned TLV:

	TLV
† Acrylonitrile	2 ppm
††Asbestos	
Amosite	0.5 fiber > 5μm/cc
Chrysotile	2 fibers > 5μm/cc
Crocidolite	0.2 fiber > 5μm/cc
Other forms	2 fibers > 5μm/cc
bis (Chloromethyl) ether	0.001 ppm
Chromite ore processing (chromate)	0.05 mg/m³, as Cr
Nickel sulfide roasting, fume & dust	1.0 mg/m³, as Ni
Coal tar pitch volatiles	0.2 mg/m³, as benzene solubles
†† Vinyl chloride	5 ppm

* * *

A1b. *Human Carcinogens*. Substances, or substances associated with industrial processes, recognized to have carcinogenic potential without an assigned TLV:

**. Acrylonitrile
4-Aminodiphenyl (p-Xenylamine)
† Benzidine — Skin
** Chloromethyl methyl ether
** Ethylene dibromide
β-Naphthylamine
4-Nitrodiphenyl

†1980 Addition.
††1980 Adoption.
**See Notice of Intended Changes.

For the substances in 1b, no exposure or contact by any route — respiratory, skin or oral, as detected by the most sensitive methods — shall be permitted. The worker should be properly equipped to insure virtually no contact with the carcinogen.

A2. *Industrial Substances Suspect of Carcinogenic Potential for MAN*. Chemical substances or substances associated with industrial processes, which are suspect of inducing cancer, based on either (1) limited epidemiologic evidence, exclusive of clinical reports of single cases, or (2) demonstration of carcinogenesis in one or more animal species by appropriate methods.

3-Amino 1, 2, 4-triazole	
†† Antimony trioxide production*	——
†† Arsenic trioxide production	——
Benzene	10 ppm
Benzo(a)pyrene	——
Beryllium	2.0 μg/m³
** Cadmium oxide production	——
Carbon tetrachloride	5 ppm
Chloroform	10 ppm
† Chloromethyl methyl ether	——
Chromates of lead and zinc, as Cr	0.05 mg/m³
3, 3'-Dichlorobenzidine	——
Dimethylcarbamyl chloride	——
1, 1-Dimethyl hydrazine	0.5 ppm
Dimethyl sulfate — Skin	0.1 ppm
† Ethylene dibromide — Skin	——
† Hexachlorobutadiene	0.02 ppm
Hexamethyl phosphoramide — Skin	——
Hydrazine	0.1 ppm
4, 4'-Methylene bis (2-chloroaniline) — Skin	0.02 ppm
Methyl hydrazine	0.2 ppm
Methyl iodide	2 ppm
††C 2-Nitropropane	25 ppm
n-Nitrosodimethylamine	——
n-Phenyl-beta-naphthylamine	——

*Cigarette smoking can enhance the incidence of respiratory cancers from this or others of these substances or processes.
†1980 Addition.
††1980 Adoption.
**See Notice of Intended Changes.

Propane sultone	——
beta-Propiolactone	——
† Propylene imine — Skin	2 ppm
† o-Tolidine	——
†† Vinyl bromide	5 ppm
Vinyl cyclohexene dioxide	10 ppm

For the above, worker exposure by all routes should be carefully controlled to levels consistent with the animal and human experience data (see Documentation), including those substances with a listed TLV.

THE COMMITTEE GUIDELINES FOR CLASSIFICATION OF EXPERIMENTAL ANIMAL CARCINOGENS

The following guidelines are offered in the present state of knowledge as an aid in classifying substances in the occupational environment found to be carcinogenic in experimental animals. A need was felt by the Threshold Limits Committee for such a classification in order to take the first step in developing an appropriate TLV for occupational exposure.

Determination of Approximate Threshold of Response Requirement. In order to determine in which category to classify an experimental carcinogen for the purpose of assigning an industrial air limit (TLV), an approximate threshold of neoplastic response must be determined. Because of practical experimental difficulties, a precisely defined threshold cannot be attained. For the purposes of standard-setting, this is of little moment, as an appropriate risk, or safety, factor can be applied to the approximate threshold, the magnitude of which is dependent on the degree of potency of the carcinogenic response.

To obtain the best 'practical' threshold of neoplastic response, dosage decrements should be less than logarithmic. This becomes particularly important at levels greater than 10 ppm (or corresponding mg/m³). Accordingly, after a range-finding determination has been made by logarithmic decreases, two additional dosage levels are required within the levels of "effect" and "no effect" to approximate the true threshold of neoplastic response.

†1980 Addition.
††1980 Adoption.

The second step should attempt to establish a metabolic relationship between animal and man for the particular substance found carcinogenic in animals. If the metabolic pathways are found comparable, the substance should be classed highly suspect as a carcinogen for man. If no such relation is found, the substance should remain listed as an experimental animal carcino gen until evidence to the contrary is found.

Proposed Classification of Experimental Animal Carcinogens. Substances occurring in the occupational environment found carcinogenic for animals may be grouped into three classes, those of high, intermediate and low potency. In evaluating the incidence of animal cancers, significant incidence of cancer is defined as a neoplastic response which represents, in the judgment of the Committee, a significant excess of cancers above that occurring in negative controls.

EXCEPTIONS: No substance is to be considered an occupational carcinogen of any practical significance which reacts by the respiratory route at or above 1000 mg/m³ for the mouse, 2000 mg/m³ for the rat; by the dermal route, at or above 1500 mg/kg for the mouse, 3000 mg/kg for the rat; by the gastrointestinal route at or above 500 mg/kg/d for a lifetime, equivalent to about 100 g T.D. for the rat, 10g T.D. for the mouse.

These dosage limitations exclude such substances as dioxane and trichlorethylene from consideration as carcinogens.

Examples: Dioxane — rats, hepatocellular and nasal tumors from 1015 mg/kg/d, oral
Trichloroethylene — female mice, tumors (30/98 @ 900 mg/kg/d), oral

Industrial Substances of High Carcinogenic Potency in Experimental Animals.

1. A substance to qualify as a carcinogen of high potency must fulfill one of the three following conditions in two animal species:

1a. *Respiratory*. Elicit cancer from (1) dosages below 1 mg/m³ (or equivalent ppm) via the respiratory tract in 6- 7-hour daily repeated inhalation exposures throughout lifetime; or (2) from a single intratracheally administered dose not exceeding 1 mg of particulate, or liquid, per 100 ml or less of animal minute respiratory volume;

Examples: bis-Chloromethyl ether, malignant tumors, rats, @ 0.47 mg/m^3 (0.1 ppm) in 2 years;

Hexamethyl phosphoramide, nasal squamous cell carcinoma, rats, @ 0.05 ppm, in 13 months

OR

1b. *Dermal*. Elicit cancer within 20 weeks by skin-painting, twice weekly at 2 mg/kg body weight or less per application for a total dose equal to or less than 1.5 mg, in a biologically inert vehicle;

Examples: 7, 12-Dimethylbenz(a)anthracene — skin tumors @ 0.12-0.8 mg T.D. in four weeks

Benzo(a)pyrene, mice 12 μg, 3X/wk for 18 mos. T.D. 2.6 mg, 90.9% skin tumors

OR

1c. *Gastrointestinal*. Elicit cancer by daily intake via the gastrointestinal tract, within six months, with a six-month holding period, at a dosage below 1 mg/kg body weight per day; total dose, rat, $\leq$ 50 mg; mouse, $\leq$ 3.5 mg;

Examples: 7, 12-Dimethylbenz(a)anthracene — mammary tumors from 10 mg 1X

3-Methylcholanthrene — Tumors @ 3 sites from 8 mg in 89 weeks

Benzo(a)pyrene, mice, 3.9% leukemias, from 30 mg T.D. 198 days

2. Elicit cancer by all three routes in at least two animal species at dose levels prescribed for high or intermediate potency.

44

Industrial Substances of Intermediate Carcinogenic Potency in Experimental Animals.

To qualify as a carcinogen of intermediate potency, a substance should elicit cancer in two animal species at dosages intermediate between those described in A3a and A3c by two routes of administration.

Example: Carbamic acid ethyl ester

Dermal, mammary tumors, mice, 100%, 63 weeks, 500–1400 mg T.D. Gastrointestinal, various type tumors, mice 42 weeks, 320 mg T.D.

Gastrointestinal, various type tumors, rats, 60 weeks, 110–930 mg T.D.

Industrial Substances of Low Carcinogenic Potency in Experimental Animals.

To qualify as a carcinogen of low potency, a substance should elicit cancer in one animal species by any *one* of three routes of administration at the following prescribed dosages and conditions:

1a. *Respiratory*. Elicit cancer from (1) dosages greater than 10 mg/m^3 (or equivalent ppm) via the respiratory tract in 6- 7-hour, daily repeated inhalation exposures, for 12 months' exposure and 12 months' observation period; or (2) from intratracheally administered dosages totaling more than 10 mg of particulate or liquid per 100 ml or more of animal minute respiratory volume;

Examples: Beryl (beryllium aluminum silicate) malig. lung tumors, rats, @ 15 mg/m^3 @ 17 months

Benzidine, var. tumors, rats, 10–20 mg/m^3 @ > 13 mos.

OR

1b. *Dermal*. Elicit cancer by skin-painting of mice in twice weekly dosages of > 10 mg/kg body weight in a biologically inert vehicle for at least 75 weeks, i.e., $\geq$ 1.5g T.D.

Examples: Shale tar, mouse, 0.1 ml $\times$ 50 = 5g T.D. 59/60 skin tumors

Arsenic trioxide, man, dose unknown, but estimated to be high

45

1c. *Gastrointestinal*. Elicit cancer from daily oral dosages of 50 mg/kg/day or greater during the lifetime of the animal.

APPENDIX B
SUBSTANCES OF VARIABLE COMPOSITION

B1 *Polytetrafluoroethylene* decomposition products*. Thermal decomposition of the fluorocarbon chain in air leads to the formation of oxidized products containing carbon, fluorine and oxygen. Because these products decompose in part by hydrolysis in alkaline solution, they can be quantitatively determined in air as fluoride to provide an index of exposure. No TLV is recommended pending determination of the toxicity of the products, but air concentrations should be minimal.

B2 *Gasoline*. See Notice of Intended Change.

B3 *Welding Fumes — Total Particulate (NOC)** TLV, 5mg/m³*

Welding fumes cannot be classified simply. The composition and quantity of both are dependent on the alloy being welded and the process and electrodes used. Reliable analysis of fumes cannot be made without considering the nature of the welding process and system being examined; reactive metals and alloys such as aluminum and titanium are arc-welded in a protective, inert atmosphere such as argon. These arcs create relatively little fume, but an intense radiation which can produce ozone. Similar processes are used to arc-weld steels, also creating a relatively low level of fumes. Ferrous alloys also are arc-welded in oxidizing environments which generate considerable fume, and can produce carbon monoxide instead of ozone. Such fumes generally are composed of discreet particles of amorphous slags containing iron, manganese, silicon and other metallic constituents depending on the alloy system involved. Chromium and nickel compounds are found in fumes when stainless steels are arc-welded. Some coated and flux-cored electrodes are formulated with fluorides and the fumes associated with them can contain significantly more fluorides than oxides. Because of the above factors, arc-welding fumes frequently must be tested for individual constituents which are likely to be present to determine whether specific TLV's are exceeded. Conclusions based on total fume concentration are generally adequate if no toxic elements are present in welding rod, metal, or

*Trade Names: Algoflon, Fluon, Halon, Teflon, Tetran.
**Not otherwise classified (NOC).

46

metal coating and conditions are not conducive to the formation of toxic gases.

Most welding, even with primitive ventilation, does not produce exposures inside the welding helmet above 5 mg/m³. That which does, should be controlled.

APPENDIX C
MIXTURES

C.1 THRESHOLD LIMIT VALUES FOR MIXTURES

When two or more hazardous substances are present, their combined effect, rather than that of either individually, should be given primary consideration. In the absence of information to the contrary, the effects of the different hazards should be considered as additive. That is, if the sum of the following fractions,

$$\frac{C_1}{T_1} + \frac{C_2}{T_2} + \ldots \frac{C_n}{T_n}$$

exceeds unity, then the threshold limit of the mixture should be considered as being exceeded. C_1 indicates the observed atmospheric concentration, and T_1 the corresponding threshold limit (See Example 1A.a. and 1A.c.).

Exceptions to the above rule may be made when there is a good reason to believe that the chief effects of the different harmful substances are not in fact additive, but *independent* as when purely local effects on different organs of the body are produced by the various components of the mixture. In such cases the threshold limit ordinarily is exceeded only when at least one member of the series $\left(\frac{C_1}{T_1} + \text{or} + \frac{C_2}{T_2} \text{ etc.}\right)$ itself has a value exceeding unity (See Example 1A.c.).

Synergistic action or potentiation may occur with some combinations of atmospheric contaminants. Such cases at present must be determined individually. Potentiating or synergistic agents are not necessarily harmful by themselves. Potentiating effects of exposure to such agents by routes other than that of inhalation is also possible, e.g. imbibed alcohol and inhaled narcotic (trichloroethylene). Potentiation is characteristically exhibited at high concentrations, less probably at low.

When a given operation or process characteristically emits a number of harmful dusts, fumes, vapors or gases, it will frequently be only feasible to attempt to evaluate the hazard by measurement of a single substance. In such cases, the threshold limit used for

this substance should be reduced by a suitable factor, the magnitude of which will depend on the number, toxicity and relative quantity of the other contaminants ordinarily present.

Examples of processes which are typically associated with two or more harmful atmospheric contaminants are welding, automobile repair, blasting, painting, lacquering, certain foundry operations, diesel exhausts, etc.

C.1A Examples of THRESHOLD LIMIT VALUES FOR MIXTURES

The following formulae apply only when the components in a mixture have similar toxicologic effects; they should not be used for mixtures with widely differing reactivities, e.g. hydrogen cyanide & sulfur dioxide. In such case the formula for Independent Effects (1A.c.) should be used.

1A.a. General case, where air is analyzed for each component:

a. *Additive effects. (Note: It is essential that the atmosphere be analyzed both qualitatively and quantitatively for each component present, in order to evaluate compliance or noncompliance with this calculated TLV.)*

$$\frac{C_1}{T_1} + \frac{C_2}{T_2} + \frac{C_3}{T_3} + \ldots = 1$$

Example No. 1A.a.: Air contains 400 ppm of acetone (TLV = 1000 ppm) 150 ppm of secbutyl acetate (TLV = 200 ppm) and 100 ppm of 2-butanone (TLV = 200 ppm)

Atmospheric concentration of mixture = 400 + 150 + 100 = 650 ppm of mixture

$$\frac{400}{1000} + \frac{150}{200} + \frac{100}{200} = 0.4 + 0.75 + 0.5 = 1.65$$

Threshold Limit is exceeded.

1A.b. Special case when the source of contaminant is a liquid mixture and the atmospheric composition is *assumed* to be similar to that of the original material; e.g. on a time-weighted average exposure basis, all of the liquid (solvent) mixture eventually evaporates.

Additive effects (approximate solution)

1. The percent composition (by weight) of the liquid mixture is known, the TLVs of the constituents must be listed in mg/m³.

(Note: In order to evaluate compliance with this TLV, field sampling instruments should be calibrated, in the laboratory, for response to this specific quantitative and qualitative air-vapor mixture, and also to fractional concentrations of this mixture; e.g., 1/2 the TLV; 1/10 the TLV; 2 × the TLV; 10 × the TLV; etc.)

$$\text{TLV of mixture} = \frac{1}{\frac{f_a}{TLV_a} + \frac{f_b}{TLV_b} + \frac{f_c}{TLV_c} + \ldots \frac{f_n}{TLV_n}}$$

Example No. 1: Liquid contains (by weight)
50% heptane: TLV = 400 ppm or 1600 mg/m³
1 mg/m³ ≡ 0.25 ppm
30% methyl chloroform: TLV = 350 ppm or 1900 mg/m³
1 mg/m³ ≡ 0.28 ppm
20% perchloroethylene: TLV = 100 ppm or 670 mg/m³
1 mg/m³ ≡ 0.15 ppm

$$\text{TLV of Mixture} = \frac{1}{\frac{0.5}{1600} + \frac{0.3}{1900} + \frac{0.2}{670}}$$

$$= \frac{1}{0.00031 + 0.00016 + 0.00030}$$

$$= \frac{1}{0.00077} = 1300 \text{ mg/m}^3$$

of this mixture
50% or (1300) (0.5) = 650 mg/m³ is heptane
30% or (1300) (0.3) = 390 mg/m³ is methyl chloroform
20% or (1300) (0.2) = 260 mg/m³ is perchloroethylene

These values can be converted to ppm as follows:

heptane: 650 mg/m³ × 0.25 = 162 ppm

methyl chloroform: 390 $mg/m^3 \times 0.18 = 70$ ppm
perchloroethylene: 260 $mg/m^3 \times 0.15 = 39$ ppm

TLV of mixture = 162 + 70 + 39 = 271 ppm, or 1300 mg/m^3

1A.c. *Independent effects*.

Air contains 0.15 mg/m^3 of lead (TLV, 0.15) and 0.7 mg/m^3 of sulfuric acid (TLV, 1).

$$\frac{0.15}{0.15} = 1; \qquad \frac{0.7}{1} = 0.7$$

Threshold limit is not exceeded.

1B. **TLV for Mixtures of Mineral Dusts.**

For mixtures of biologically active mineral dusts the general formula for mixtures may be used.

For mixture containing 80% nonasbestiform talc and 20% quartz, the TLV for 100% of the mixture is given by:

$$TLV = \frac{1}{\frac{0.8}{20} + \frac{0.2}{2.7}} = 9 \text{ mppcf}$$

TLV of nonasbestiform talc (pure) = 20 mppcf
TLV of quartz (pure) =

$$\frac{300}{100 + 10} = \frac{300}{110} = 2.7 \text{ mppcf}$$

Essentially the same result will be obtained if the limit of the more (most) toxic component is used provided the effects are additive. In the above example the limit for 20% quartz is 10 mppcf.

For another mixture of 25% quartz, 25% amorphous silica and 50% talc:

25% quartz — TLV (pure) = 2.7 mppcf
25% amorphous silica — TLV (pure) = mppcf
50% talc TLV (pure) = 20 mppcf

$$TLV = \frac{1}{\frac{0.25}{2.7} + \frac{0.25}{20} + \frac{0.5}{20}} = 8 \text{ mppcf}$$

The limit for 25% quartz approximates 9 mppcf.

APPENDIX E
Some Nuisance Particulates[o]

TLV, 30 mppcf or 10mg/m^3 of total dust < 1% quartz, or, 5 mg/m^3 respirable dust

Aluminum Oxide (Al_2O_3)	Kaolin
Calcium carbonate	Limestone
Calcium silicate	Magnesite
Cellulose (paper fiber)	Marble
Portland Cement	Mineral Wool Fiber
Emery	Pentaerythritol
Glycerin Mist	Plaster of Paris
Graphite (synthetic)	Silicon
Gypsum	Silicon Carbide
Vegetable oil mists (except castor, cashew nut, or similar irritant oils)	Starch
	Sucrose
	**Tin Oxide
	Titanium Dioxide
	Zinc Stearate
	Zinc oxide dust

o) When toxic impurities are not present, e.g. quartz < 1%.

APPENDIX F
Some Simple Asphyxiants[p]

Acetylene	Hydrogen
Argon	Methane
Butane	Neon
Ethane	Propane
Ethylene	Propylene
Helium	

p) As defined on pg. 6.

APPENDIX G
Conversion of mppcf to Mass Concentration

Calculations for Conversion of Particle Count Concentration (by Standard Light Field — Midget Impinger Techniques), in mppcf, to Respirable Mass Concentration (by Respirable Sampler) in mg/m^3.†

1. In 1967, Jacobsen and Tomb,† derived an empirical relationship of 5.6 mppcf to 1 milligram of respirable dust per cubic meter of air, based on 23 sets of samples, mostly coal dust. The following calculation results in an equivalence of 6.37 mppcf to 1 mg/m^3 of respirable dust. Thus; an approximate ratio of 6 mppcf to 1 mg/m^3 of respirable dust is suggested for conversion of TLVs from a count to a mass basis when the density and mass median diameter have not been determined.

†"Relationship Between Gravimetric Respirable Dust Concentration and Midget Impinger Number Concentration," by Murray Jacobson and T. F. Tomb, AIHAJ, 28: Nov.–Dec. 1967.
**See Notice of Intended Changes.

2. **Basic assumptions:**

a) **Average density for silica containing dusts ≈ 2.5 gms/cm^3 (2500 mg/cm^3). Pulmonary significant dust densities may vary from 1.2 gm/cm^3 for coal dust to 3.1 gms/cm^3 for Portland Cement. Silica densities vary from 2.2 (amorphous) to 2.3 (cristobalite and tridymite) to 2.5 (alpha-quartz.) gms per cm^3.**

b) **The mass median diameter (mmd) of particles collected in midget impinger samplers and counted by the standard light field technique, *and* collected in a respirable sampler is approximately 1.5 μm or 1.5×10^{-4} cm. This assumption is, of course, quite arbitrary since the mmd of all dust clouds is quite variable, depending on many independent parameters, such as source of dust, age of dust cloud, meteorological conditions, etc.**

3. Calculation:

a) vol. per particle: $4/3\ \pi r^3$; $r = 0.75 \times 10^{-4}$ cm
$= 4/3 \cdot \pi \cdot (0.75 \times 10^{-4})^3$
$= 1.77 \times 10^{-12}\ cm^3$

b) wt. per particle = vol. × density
$= 1.77 \times 10^{-12}\ cm^3 \times 2.5 \times 10^3\ mg/cm^3$
$= 4.425 \times 10^{-9}$ mg/particle

c) 1 particle/$ft.^3$ = 35.5 part./m^3
(since 35.5 cu ft = 1 cu m.)
10^6 part./ft^3 = mppcf = 35.5×10^6 part./m^3

wt. of 1 mppcf ≡ 35.5×10^6 part./m^3 × 4.425×10^{-9} mg/part.

1 mppcf ≡ 0.157 mg/m^3
or
6.37 mppcf ≡ 1 mg/m^3
or approximately 6 mpccf ≡ 1 mg/m^3.

4. Equivalent TLVs in mppcf and mg/m³ (respirable mass) for Mineral Dusts.

Substance	Threshold Limit Value		
	Count mppcf	Resp. Mass mg/m³	Total Mass* mg/m³
Silica (SiO_2)			
Amorphous	20	(3)**	(6)
Cristobalite	1.5	0.05	0.15
Fused silica	3	0.1	0.3
Quartz	3	0.1	0.3
Tridymite	1.5	0.05	0.15
Coal Dust	(12)	2	(4)
Diatomaceous earth, natural	—	1.5	—
Graphite (natural)	15	(2.5)	(5)
Mica	20	(3)	(6)
Mineral wool fiber	—	(5)	10
Nuisance particulates	30	(5)	10
Perlite	30	(5)	(10)
Portland Cement	30	(5)	(10)
Soapstone	20	(3)	(6)
Tripoli	(3)	0.1	(0.3)

*Unless otherwise specified, respirable mass is presumed to equal approximately 50% of total mass.

**All values in parentheses ( ) represent newly calculated values based on equivalence of 6 mppcf = 1 mg/m³ respirable mass and respirable mass = 50% total mass.

Glossary of Acronyms Used in This Report

ACS American Chemical Society
ACGIH American Conference of Governmental Industrial Hygienists
ANSI American National Standards Institute
CFR Code of Federal Regulations
EPA U.S. Environmental Protection Agency
FDA U.S. Food and Drug Administration
HEPA High-efficiency particulate air (filter)
IDLH Immediately dangerous to life and health
LC_{50} Concentration of material in air that causes death of 50% of test animals
LD_{50} Single-dose quantity of material that will cause death of 50% of test animals.
MSHA U.S. Mine Safety and Health Administration
NFPA National Fire Protection Agency
NIOSH National Institute for Occupational Safety and Health
OSHA U.S. Occupational Safety and Health Adminstration
PEL Permissible Exposure Limit
PMR Proportionate Mortality Ratio
RCRA Resource Conservation and Recovery Act of 1976
SIR Standardized Incidence Ratio
SMR Standardized Mortality Ratio
TLV® Threshold Limit Value established by ACGIH
TWA Time-Weighted Average

Index

Notes